PHOTOVOLTAIC PRODUCT DIRECTORY & BUYERS GUIDE

Home, Farm, Industrial, and Institutional Systems, Components and Appliances

R.L. Watts S.A. Smith
R.P. Mazzucchi

 VAN NOSTRAND REINHOLD COMPANY
NEW YORK CINCINNATI TORONTO LONDON MELBOURNE

TK
1056
·W38
1982

Revised edition first published in 1982
Copyright © 1982 by R.L. Watts, S.A. Smith, and R.P. Mazzucchi
Library of Congress Catalog Card Number 81-71474
ISBN 0-442-29239-2

Printed in the United States of America

Portions of this work were prepared for the U.S. Department
of Energy under contract DE-AC06-76RLO 1830
Van Nostrand Reinhold Company Inc.
135 West 50th Street, New York, NY 10020

Van Nostrand Reinhold Publishers
1410 Birchmount Road, Scarborough, Ontario MIP 2E7

Van Nostrand Reinhold Australia Pty. Ltd.
17 Queen Street, Mitcham, Victoria 3132

Van Nostrand Reinhold Company Ltd.
Molly Millars Lane, Wokingham, Berkshire, England RG11 2PY

16 15 14 13 12 11 10 9 8 7 6 5 4 3 2 1

ACKNOWLEDGMENTS

The authors wish to acknowledge the support, encouragement and contributions of a number of individuals and organizations associated with the Department of Energy Photovoltaic Program and the photovoltaics industry.

Department of Energy ideas and funding were contributed by the staff of the Photovoltaics Division including Andrew Krantz, Elaine Smith and Paul Maycock. Planning assistance and support were given by George Tenet, Ray Kendall and the PV executive council of the Photovoltaics Division of the Solar Energy Industries Association.

We would also like to express our appreciation to the late John Hesse and the Jet Propulsion Laboratory staff, John Bozek of NASA Lewis Research Center, and to Robert Koontz, Tom West, and Peter Thompson of SERI for reviewing the draft.

We would also like to thank Roger Baird, Clem Bloomster, Ted Blumenstock, Bob Brown, Bill Brussear, Richard Campbell, Molly Craver, Jim Cullen, Dennis Delans, John Eckel, Candy Eliason, Rita Harrington, Mike Keeling, Brian Kennedy, Carl Kotila, Bill Lamb, Ernie Lampert, Jo-Mari Lett, Roger Little, Vic Luke, Bob McGinnis, Stephen Madigan, Jim Marler, Jim Merril, Mike Nelson, Jim Padula, K. V. Ravi, Vince Rice, Rob Robinson, Rick Scott, Ishaq Sharayer, Jim Skane, Diane Smith, Bob Stewart, Ron Strathman, Sam Taylor, Mary Trigg, Fritz Wald, Bob Walker, Vern Weekman, Art White, Tom Wilber, Robert Willis, John Wurmser, and Jack Zohar.

A PHOTOVOLTAIC PRODUCT DIRECTORY AND BUYER'S GUIDE

PREFACE

Photovoltaic (PV) generation of electricity marks the dawn of a new energy era. In total silence PV systems remarkably and simply convert light directly into electricity. This electricity powers the tools of civilization from the mountain tops to the seas and into the far reaches of outer space. It is now possible to build a home anywhere you desire or even travel to the moon and have the conveniences of a secure and reliable PV power supply. In recognition of the energy savings potential of PV systems for terrestrial applications, the Department of Energy in conjunction with industry has conducted research and development to accelerate price reductions and product adaptation. This effort has succeeded brilliantly, bringing this space age dream down to earth so that in many applications it is now cost effective to use PV energy supplies.

Photovoltaics are providing energy for an ever growing number of applications worldwide. More than ten thousand residences now use PV power to provide for daily electrical needs. Vacation chalets and remote residences enjoy the advantages of simple and easily maintained PV power supplies. More than ten thousand navigational beacons and warning lights, using stored photovoltaic power, provide safe passage for ships and aircraft. The list of viable applications continues to grow as inventive thinkers ponder the possibilities of PV technology. Although it is not yet cost-effective to use PV electricity as if it were provided by a central utility, there are many applications where PV is now the most cost effective and reliable power supply. In areas where utility lines are difficult to access or where energy management and lifestyle changes conserve electricity, PV systems are viable today. PV systems are also particularly useful for applications where portability is desired or where energy needs coincide with solar intensity for such applications as ventilating and water pumping.

The information presented in this directory should be of interest to a wide audience, because it describes products and appliances that can be used in or on the home, farm, or factory as well as on the road or seas and in remote areas of the great outdoors. Reliable and cost-effective PV powered products are available for those who are willing to seek them out and apply them. This directory is intended to help you decide if currently marketed photovoltaic products are appropriate for your needs.

CONTENTS

1.0 INTRODUCTION

A major obstacle to the widespread acceptance and use of photovoltaics (PV) has been the lack of practical information on the way these systems work, and the types of systems and products available in 1981. The purpose of this directory is to provide up-to-date information on the PV industry in a form that you can easily use to become better acquainted with PV. The directory also includes a list as complete as possible of the sources of PV products as of 1981. The directory/buyer's guide includes material that explains the language (used in the industry), operation, design guidelines, and approximate list prices of currently available PV systems and their components.

The directory/buyer's guide is intended to be a comprehensive compilation of PV products, applications, and manufacturers. The focus is to identify and discuss PV products and system components that can be purchased off-the-shelf or ordered from catalogs that are available as of 1981. The discussions are written from your perspective as a potential buyer and are intended to reveal products that can meet your needs. The information should also be of value to manufacturers because it identifies opportunities to meet developing market demands.

The format of the directory has been selected to provide assistance for anyone interested in PV as well as provide an organization of material that can be easily referenced by professionals in the field. The directory/buyer's guide is organized to:

- help you understand photovoltaic systems (Chapter 2.0),
- show products available off-the-shelf and assist you in using them (Chapter 4.0),
- direct you to sources of expert help for the challenging applications (Chapter 5.0),
- help you determine if PV products can meet your needs (Chapter 6.0), and
- provide information on actual PV user experiences (Chapter 7.0).

We have also included four appendices to provide detailed information in the following areas:

- information on various financial incentives available from state and federal governments (Appendix I),
- sources of additional information on photovoltaics (Appendix II),
- a matrix indicating the sources of various types of PV products (Appendix III), and
- a listing of the addresses of PV product suppliers (Appendix IV).

The material presented is based upon information from publications and programs sponsored by the Department of Energy, responses to an inquiry of PV product manufacturers and publications from private entities. A literature search of material dealing with PV manufacturers, products, and projects yielded leads on information in the first and third categories. The information from the second category resulted from announcements in the trade press that publicized the compilation of the directory, and inquiries mailed to all identified entities producing photovoltaic products or products amenable to PV power supplies. Once a draft of the catalog was complete, four workshops were held in Washington D.C., Boston, Phoenix and Los Angeles, to ensure that all product information is accurate and up to date.

Some firms have undoubtedly been overlooked due to the rapid growth of the industry. Since our efforts far exceed those that most potential purchasers exercise to identify PV products, we feel justified in asserting that this directory is comprehensive. We have painstakingly attempted to contact all firms listed in directories prepared by the Solar Energy Industries Association, Department of Energy, Solar Energy Research Institute, Solar Engineering Magazine, and the Solar Age Resource Guide, among others. Firms who have been overlooked are requested to contact the authors to insure that they will be included in the next update.

All of the information on products contained in this directory has been assembled on the basis of catalogs and data sheets submitted to us by PV suppliers. Since it is not within the scope of this effort to test and verify claims by manufacturers, we recommend that purchasers obtain additional information from PV users, suppliers and other reliable sources. Reputable product suppliers frequently offer warranties and maintain lists of satisfied customers to substantiate their claims. Neither Battelle nor the government recommends the purchase or use of any specific product or supplier.

2.0 BACKGROUND INFORMATION ON ELECTRICITY AND PHOTOVOLTAICS

This chapter is intended to introduce PV technology and provide basic information useful to readers who are unfamiliar with PV energy systems. After briefly summarizing the historical development of photovoltaics, we provide a description of the building blocks of a PV system. Finally a short glossary of electrical terms is included to help you understand the operation and specification of PV products.

The derivation of the term "photovoltaic" is from two Latin words: "photo" meaning of, or produced by, light; and "voltaic" meaning of, or producing, an electric current and voltage. Photovoltaic, or solar cell, technology is an exciting alternative source of energy that converts sunlight directly into electricity without any moving parts. Photovoltaic systems operate efficiently in a wide range of applications including small, low-power devices for remote communication instruments; mid-size systems for residences and schools; and large power systems for high-demand operation.

Vigorous efforts are under way within the Department of Energy's National Photovoltaic Program to increase the number of PV systems installed across the country. These installations, which are discussed in Chapter 5, promote greater public awareness of the practicality of PV, fortify confidence in their use, and help stimulate growth of the industry. Industry growth and better manufacturing techniques have dramatically improved the economic attractiveness of PV. Continued encouragement of PV power generation will aid in the orderly transition from an economy based on fossil fuel to one founded on alternative sources of energy.

In 1954, scientists at Bell Laboratories reported an improved solar cell that enabled PV technology to be used as a practical energy source. The primary use of solar cells was to power spacecraft equipment. In 1975, ERDA began a research and development program to hasten production of solar cells and promote their use. The program has grown to include hundreds of scientists who work in industry, universities, and laboratories across the country and has brought about significant advancements in efficiency, durability, and system cost effectiveness.

Today, most of the reliable and inexpensive solar cells are made of single-crystal silicon. Silicon is the second most abundant element on earth. It is found, for example, in various kinds of sand. After an extensive purification process, silicon crystals are "grown" and then cut into slices to form the basis of the solar cells. When sunlight strikes the solar cell, internal electrons are energized and electricity is generated. Useful electricity is drawn off through wires attached to the cells.[a]

2.1 PHOTOVOLTAIC SYSTEM COMPONENTS

The following material provides brief definitions of the components in a typical PV system. Not all PV systems have or need all of these components, but they are introduced here for your information.

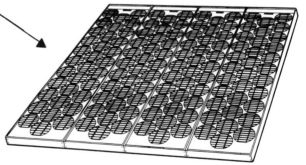

Photovoltaic Cells are discs or squares of specially treated silicon (or other material) that generate a voltage, when exposed to light.

Photovoltaic Modules contain a number of PV cells that are encapsulated (sealed) between metal and transparent plates. Modules are either flatplates or use special designs to concentrate the sunlight on the cells. The electrical output of the modules depends on the efficiency and number of cells.

A Photovoltaic Array is a group of PV modules connected together to produce a predetermined voltage for either energy storage or to be used directly (without energy storage) by a direct current DC appliance.

---

[a] Solar Energy Research Institute 1980. Photovoltaics Solar Electric Power Systems. SERI/SP-433-487. Golden, Colorado.

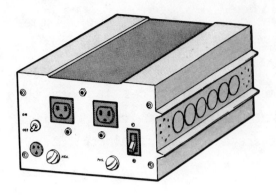

Inverters change the direct current (DC) produced by the cells or stored in the batteries to alternating current (AC) for use with AC products and appliances.

Charge Controllers protect the battery system from both excessive charge and discharge.

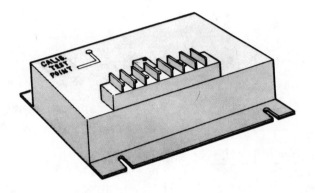

Energy Storage is usually done with lead acid batteries. The type and size depend on the requirements of the specific application.

End-Use equipment (i.e., load) can be selected from a wide variety of AC and DC products including fluorescent lights, refrigerators, pumps, fans, etc.

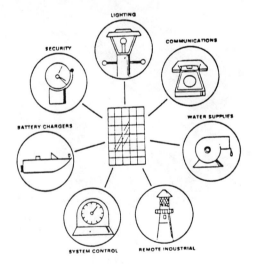

2.2 ELECTRICAL DEFINITIONS

If you are the type of person who frequently reads Electrical World, and knows the difference between ampere, volt, watt, power and energy, you can skip this section. In this section we provide definitions of some of the basic elements of electricity. This information will help you in reading catalogs supplied by PV manufacturers/distributors, and it will help you when expressing your energy needs to these people.

Ampere (A) is a standard unit of current. It is the rate of flow of a charge past a given point in an electrical circuit. This is analogous to the quantity of water passing by a given point in a garden hose. The two types of electrical current are direct (DC) and alternating (AC). It is not really that important to know the specific details of these currents, only that some products/appliances operate strictly on DC or AC and some, such as portable TV sets, and certain universal (brush type) motors can run on either.

Lumen (lm) is a measurement of the amount of light produced by an incandescent or fluorescent light bulb. A 100 W incandescent light bulb produces about 2000 lumens of light.

Voltage (V) is a measurement of potential difference between two points or an electromotive force. In layman's terms it is the force that causes the current to flow through a wire. Most automobile batteries are either 6 or 12 volts.

Watt (W) is a unit of power. One watt is the amount of power that is supplied when a current of one ampere is driven by a potential difference (i.e., force) of one volt. If you look at a typical light bulb, it will state that it requires 25, 50, or 100 W of power to operate. A watt is not a measurement of the amount of light given off by a light bulb.

Watt-Hour (Wh) is a unit of energy measured as the product of power (i.e., one watt) and time (one hour). If a 25 W light bulb is operated for one hour, it will use 25 Wh of electricity.

Ampere-Hour (Ah) is sometimes used as a unit of energy where voltage is fairly constant. For instance, storage batteries have a fairly constant voltage and are frequently rated in terms of ampere hours of capacity. Similarly, an electric clock requires a fairly constant current and its energy requirements can also be specified in Ah. Ampere-hours can be changed to watt hours by multiplying by the voltage of the system. In a 6 V system with 2 Ah of capacity, the total would be 6x2 = 12 WATT HOURS of energy. In a 12 V system with 2 Ah of capacity, there would be 24 WATT HOURS of energy.

3.0 READY TO USE PV PRODUCTS AND SYSTEMS

This chapter is written for people who want the conveniences of modern electric living and a) own recreational vehicles or water craft, b) own a remote retreat home or live in a locality where there is no dependable utility, or c) who would like to achieve a measure of energy independence using a renewable energy resource. It is also written for farmers, industrialists, and construction workers who wish to use electrical or electronic equipment away from power lines. One can watch television, read at night, wash clothes, drink pure water, keep food and medicines fresh, and stay cool in hot weather by using products that are powered by photovoltaics without dependence on an electric utility system. Remote communications, corrosion protection, crop irrigation and stock watering are other examples of modern PV use.

The following sections give an idea of the various types of PV products and systems currently available off-the-shelf. These PV systems are pre-engineered to meet specific end uses and include all the components necessary to be placed into immediate service. The PV systems have been sorted into five categories which are: Products Without Batteries, Battery Chargers, Power Packages, Home Electric Systems, and Partial Systems. For each we discuss typical applications and list the available products. Approximate prices are often given but in an age of continuing inflation, they can quickly be out of date.

3.1 PRODUCTS WITHOUT BATTERIES

Some electrical needs occur, or can be shifted to occur, only when the sun shines. Such applications are termed "sun synchronous" and do not require the use of storage batteries. Energy for cooling work and living areas, and irrigation pumping are ideal candidates for PV power supplies. Such items as attic fans, casa blanca fans, evaporative coolers for RV's and residences, circulating pumps for solar hot water heaters, and domestic or agricultural water pumps are available with PV power supplies.

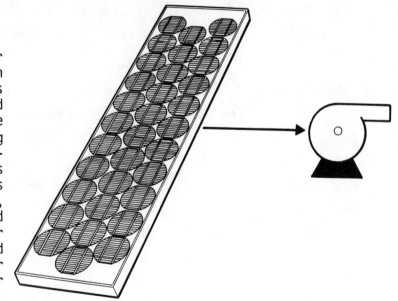

These are efficient PV applications because the output of the arrays is immediately utilized to the desired end. For example, PV powered exhaust fans installed in buildings, vehicles, and boats are used to prevent overheating -- the more intensely the sun shines, the more rapidly the fan removes excess heat. Furthermore, these systems are the most maintenance free and long-lived because of their inherent simplicity and freedom from electical storage requirements. Table 3.1 lists some of the commercial fans available.

TABLE 3.1. PV Powered Fans and Vents

Item Catalog No.	Description	Company	Rating	Approx. List Price
Solar Vent "SV"	Replacement for 12" Wind Turbine Roof Vent	Braden Wire & Metal Products	NA*	$225.00
Parmak Solar Vent Fan	Attic Fan	Parker McCrory	NA*	$298.95
Solar Breeze	12" diameter Fan with Matched PV Module	Photowatt	500 CFM	
Solar-Turbin	Sun or wind powered Ventilation Fan	Photowatt	1400 CFM	
Solar Fan	Solar Powered Attic Fan #138	Solar Thermal Inc.	800-1400 CFM in attic use	
The Solarvent	Sunpowered Ventilating Fan. 5-Blade 8" Diameter Aluminum Fan	Solarex Corp.	250 CFM 6 Watt 4-60 Unipanel	$299.95
Solar Attic Fan PVF 1204A	3-Blade Fan for Vent Use in Buildings & Attics	United Energy Corp.	3.6 Watts	
Solar Ceiling Fan PVF 1212C	3-Blade Fan of "Casa Blanca" Type	United Energy Corporation	20.3 Watts	
Casa Blanca Fan	3 blade "Casa Blance"	WM Lamb Co. & Energy House	2000 cfm	$465

* Not available.

One attractive feature of these fans is that they can be retrofitted to existing buildings by the handyman without shock or fire hazard, often at a lower cost than connecting to the building power supply. In addition, controllers are unnecessary since the array automatically fulfills the control requirements. Thus, they are frequently installed in houses that actually have utility power available.

Another product contributing to living comfort is the photovoltaic-powered evaporative cooler for cooling recreational vehicles and residences. The three sizes presently available with matched PV power supplies are listed in Table 3.2.

A PV powered circulating pump for a conventional solar collector can improve system performance and reliability. It does this by circulating the fluid rapidly through the collector, when the sun is at full brightness, and pumps slowly, when the sun is partly obscured. This "proportional control" is said to enhance the efficiency of a solar water heating system by as much as 15% above that possible from conventional systems that use household current and temperature sensitive instruments to provide system control.

TABLE 3.2 Evaporative Coolers with Matched PV Power Supplies

Item	Description	Company	Rating	Approx. List Price
RV Cooler	LA-8 RV Evaporator Cooler with 2 ASI 16-2000 panels	Wm Lamb Co. & Energy House	600 CFM	$1470
Evaporator Cooler	LA-9 Residential Evaporative Cooler with 2 ASI-16-2000 Solar Panels	Wm Lamb Co. & Energy House	1200 CFM	$1375
Residential Evaporative Cooler	Residential Evaporative Cooler Using DC Motors with 450 W of ASI 16-2000 Solar Panels	Wm Lamb Co. & Energy House	5000 CFM 1/2 hp fan Motor With Circulating Pump	$7200

Some pumps available for this purpose with matched PV modules are listed in Table 3.3.

Pumping water for irrigation, into storage tanks or directly onto the land are other practical sun synchronous applications. Systems that are available from Applied Solar Energy/Solar Electric International and Rainbird are shown in Table 3.4.

An additional system is available from Energy House and Wm. Lamb Co. for deep well pumping. This unit is designed for residential and stock watering applications or for providing water for campsites and golf courses. It uses the tested lift jack principle used on windmills for decades with a 1/2 hp DC motor located on the ground surface. The design is adaptable to lifts to 80'

TABLE 3.3 Low Flow Rate PV Pump Systems

Item & Catalog No.	Description	Company	Rating	Approx. List Price
Circulating Pumps Model 1500 & 8200	DC pump w/matching PV array	A.Y. McDonald Mfg. Co.	200 Watts and larger pumps	$2500 and up
PV Power Pak	1-30 Watt Panel & Circulating Pump	Mid Pacific Solar Corp.	33 Watt Pump	$500
PV Power Pak #1000	1-20 Watt Panel & Circulating Pump	Solar Disc	20 Watt Panel & Circulating Pump	$400
PV Power Pak #1500	1-30 Watt Panel & Circulating Pump	Solar Disc	30 Watt Panel & Circulating Pump	$500
Hot Water Circulator 12 DC-2	2-20 Watt #1212 PV Modules	United Energy Corp.	11 ft of head 5 gpm	
Hot Water Circulator 24 DC 2	2-20 W #1212 PV Modules	United Energy Corp.	11 ft of head 5 gpm	
Hot Water Circulator #12 OC1	1-20 Watt #1212 PV Module & Circulating Pump	United Energy Corp.	4 ft of head 3 gpm	

Name of Product	Specifications	Approx. Price
A. Y. McDonald Remote pumping system. Models 8200 & 1500 for high capacity & high heat applications.	Single & dual impeller, self-priming centrifugal pumping systems using DC motors. Systems connect directly to solar array or battery.	$4,900-$20,000 (with array, support hardware and wiring).
SEI-Sun Pump Model 250M (medium lift: 2.5 lt/sec @ 4.5m) Model 250L (low lift: 6 lt/sec @ 2m)	The lightweight, portable Sun Pump is a self-contained system comprising a 250-watt PV array, a pumpset, a Maximum Power Controller (MPC), and a mounting frame. It is particularly well suited for use in rural areas and other places not served by an electric grid.	$6,250 for single units; quantity discounts available
Emergency Water Supply Package	Designed to provide safe drinking water for up to 5,000 people from lake, stream, or open well. The system includes a PV-powered pump, a chlorinator and 2 25,000-litre collapsible storage tanks.	$10,000
Rain Bird Irrigation Pumping System Model CSB 33B for low head application	1/3 HP single stage self priming centrifugal pumping system with DC motor & long life brushes 396 W_p solar array. Incl. batteries, controls support hdwe. & wiring. @ 15' - 20' Head Typical 30 - 45 GPM capacity	$8,115 To Delete Batteries & Controls order CSB 33D @ $5,691
Irrigation Pumping System CSB 50B for medium head applications	1/2 HP single stage self priming centrifugal system WDC motor & long life brushes, 528W_p solar array, Incl. batteries controls support hdwe. & wiring. @ 20' - 30' Head. 35 - 45 GPM	$9,480.75 To Delete Batteries & Controls Order CSB 50D @ $7,326.75.
Irrigation Pumping System CSB 100B for medium head applications	1 HP single stage self priming centrifugal system with DC motor long life brushes 1,056 W_p. Array incl. batteries, controls, support hdwe. & wiring.	$16,962.75 To Delete Batteries & Controls Order CSB 100D @ 13,929.75
CSB 150B	D.O. Exc. 1 1/2 HP and 1584 W_p array.	$22,917 To Delete Batteries & Controls Order CSB 150D @ 19,884
CSB 200B	D.O. Exc. 2.0 HP $2,112 W_p array @ 40' - 50' Head 45 - 90 GPM	$30,035.25 To Delete Batteries & Controls. CSB 200D @ $26,123.25

or less and up to about 380' with correspondingly reduced flow (the greater lift is accompanied by reduced flow). Contact Wm Lamb or Energy House for full details on this system that sells for about $6,650.

The TriSolarCorp makes a variety of custom designed pumping systems. Their pumping systems utilize a special power matching circuit with improved pumping efficiency in the morning and afternoon.

3.2 BATTERY CHARGERS

Sun-powered battery chargers can be added to an existing battery powered appliance or system to provide additional power and to maintain batteries at a high charge level. PV powered battery chargers are available in a variety of sizes and voltages to match the battery. For as little as $20 one can purchase a charger for 3, 6 or 9 volt nickel cadmium (flashlight size) batteries. Much larger systems are available for industrial or recreational use. Currently marketed battery charger systems are listed in Table 3.5.

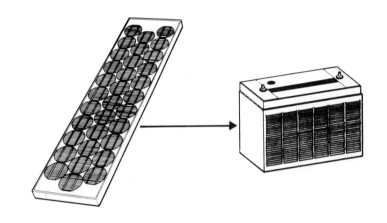

In instances where a backup generator is available, it isn't absolutely essential to precisely size the PV battery charger. You can start with a small PV unit and add more modules if desired. One caution is in order - you may save money by buying one charge controller large enough to handle the final complement of modules desired. Adding PV modules would then be fairly simple as long as the rated capacity of the charge controller and wiring is not exceeded, and the batteries charging rate is not exceeded.

TABLE 3.5. PV Powered Battery Chargers

Application	Supplier Description	Catalog Number	Rating	Charge Controller	Approx. $ List*
Rechargeable Flashlight	Energy House	Sunstor Jr.	AA-3-5 hr C 6-10 hr D 8-12 hr	Not Needed	$12.50
Nicad Pack	Solarex, Multivolt	5520400	50 ma @ 3, 6 or 9V	Not Needed	$20
Boats & RVs	Solarex, Marine	1270 MP HE51MP	0.65A @ 12V/2.1A @14V	Included	$445
"	Free Energy Systems Marine can be walked on when mounted	12-33 SL 129 SL	2.0A @ 14V/0.6A @ 14V	BVR 12-2 BVR 12-07	
"	United Energy Corporation	1204 1212	0.3 @ 12V 1.1A @ 12V	Not Needed? Not Needed?	
Marine Environment	Tideland Signal Hermetically sealed in glass	MG 600	0.6A @ 6V	Not Needed with pure lead batteries	

*For panel including blocking diode and charge controller if needed.

3.3 PV POWER PACKAGES

PV power packages consist of panels, storage batteries, and power conditioning equipment. These systems are available from a variety of sources in many sizes. Their performance depends upon the type of electrical needs and the amount of sunshine available.

Two basic types of power packages are available -- those that provide direct current (DC) only and those that also provide alternating current (AC). The major difference between the systems is the presence of an "inverter" that converts DC into AC power. Because of the power requirement of the inverter, AC systems are significantly less efficient than their DC counterparts. For more detailed information on inverter efficiencies the reader is directed to Section 4.5.

When selecting a PV power package there are several key issues to consider in addition to determining if AC or DC power is needed. These include:

- the expected peak load
- the average energy requirements
- the expected climate conditions
- the required system reliability

The storage and power conditioning systems must be large enough to handle the maximum or peak load. In the case of AC power, this usually occurs when induction motors are started -- check the "surge" power requirements. In the case of DC power, it is the sum of the appliance ratings that will be used at the same time. Not only must this peak be met by storage, but the daily and seasonal average power requirements must be considered to insure that the batteries are maintained at a satisfactory level of charge.

Two climatic factors are of primary importance; the expected levels of solar insolation (sunshine intensity), and the expected temperature extremes. Obviously the amount of sunshine available will dictate the size of the PV array. Temperature is important since it affects the performance of the array and the storage systems. If freezing conditions are expected the batteries must be suitably protected.

Finally, the issue of system reliability requirements must be considered. If power is needed for essential uses such as navigational or safety lighting, the system must be oversized to insure that sufficient power is available under any circumstances such as extended cloudy periods. PV systems for such applications should be discussed with a supplier who understands potential liability risks. On the other hand, if the system is used for recreational purposes or intermittent applications, oversizing might not be economically attractive. The sizing of PV power packages must be done properly. A report by NASA Lewis, DOE/NASA/0195-81/1 "PV Stand Alone Systems Preliminary Engineering Design Handbook" will be helpful in this task. A preliminary or "first out" sizing procedure is outlined in Chapter 4.

3.3.1 DC Power Packages

Most suppliers of PV modules have sold a sizeable number of custom designed DC power packages, but at present, only a few maintain an inventory of standard systems or send catalog sheets. Most module manufacturers will assist potential purchasers design and specify a PV system to meet personal needs. However, this section considers only those systems available off-the-shelf. The purchaser of a DC power system must keep in mind that many common AC home appliances cannot be used without purchasing a DC to AC inverter. There are, however, a growing number of DC appliances that are discussed in Section 4.2.

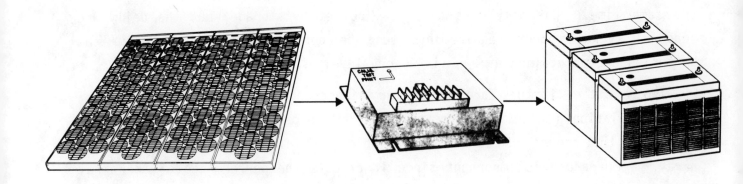

Purchasing a PV power package and using it for a variety of purposes is quite straightforward. The power packages work in any climate; however, the energy output is directly related to the amount of sunshine striking the panels. Seasonally, the systems can supply more power in the summer than in the winter in northern climates. As long as the electrical needs are adjustable and the user can tolerate reduced availabilities during periods of low insolation, the performance information supplied by the system manufacturer may be sufficient to size a satisfactory system.

These systems _are_ intended to supply power where it is inconvenient or excessively expensive to make a connection to a utility grid. Utility lines to connect an individual user usually cost in excess of $10,000 per mile and are often maintained at the purchaser's expense. With the current tax credit situation, it is possible to install a modest sized solar electric generator system and have no subsequent utility bills for a comparable investment. It is particularly attractive to install a modest system and add capacity each year while the tax credit lasts. The economic benefits of such a strategy is discussed in detail in Section 6.4, and additional information on state tax credits is provided in Appendix I.

These power packages are rated in their ability to satisfy a given load measured in ampere hours (or watt hours) per day or week. Guidance in estimating your anticipated power needs is provided in Section 4.1. Most of the currently available PV power packages are rated far below the average energy requirements for conventional residential applications. This is a consequence of the currently relatively high price of PV power compared to that available from utility grids. It makes more economic sense to make expenditures to minimize your electrical needs at first than to purchase huge PV arrays initially. Table 3.6 lists some DC power packages currently available.

Marine power systems are used in increasing numbers for critical applications, other than the marine environment for which they were specifically designed, because of their reputation for higher reliability. Because of strong competition for PV system sales, the buyer can get what he's paying for in most instances. Purchasing premium quality components and professional design services are an absolute must for applications where human life is at stake or liability of property loss exists.

3.3.2 AC Power Packages

With an adequately sized AC power package one can operate conventional household appliances. Special AC power packages can be designed to connect to the electric utility supply lines so that excess power can be sold to the utility (with utility approval). These will become more common in the future, but at the present time they are relatively rare. The systems listed in this section are not intended to connect to conventional AC power grids.

An AC power package typically costs more than DC counterparts for three reasons: a) there is the added cost of an inverter, b) PV array sizes and battery storage must be increased 15% or more to make up for inverter losses, and c) typical AC appliances have been designed with little regard for

3.10

TABLE 3.6. DC Power Package

Application	Description	Supplier/Mfg.	Typical Average Rating	Approx. List Price
Home Electric Power System with Modules Mounted Outdoors.	4-33 Watt PV Modules (Model ASI 16 -2000). 3 batteries 105AH, 12 V each (Delco 2000). 1-Control Panel 1-Pkg. Mounting Hardware.	Arco Solar	44AH/Day	$2500
Home Electric Power Systems with Modules and Batteries Outdoors. Controls are indoors. (Model P-1-12)	1-33 Watt PV Module (Model ASI 16 -2000). 1 Battery 105 AH, 12V (Delco 2000). 1 Control Panel 1 Set Cables & Wires	Solarwest Electric	7.5-13AH/Day	$900
Model P-2-12	Same as above but includes: 2 PV Modules 2 Batteries	Solarwest Electric	15-27 AH/Day	$1550
Model P-3-12	Same as above but includes: 3 PV modules 2 batteries	Solarwest Electric	22.5-40 AH/Day	$2200
Model H-4-12	Same as above but includes 4 PV Modules 3 batteries	Solarwest Electric	30-53 AH/Day	$2650
Industrial Grade Systems for High Reliability Uses.	185 Standard Solar Generators for High Risk Applications.	Tideland Signal Corp.	2 AH/Day @ 6 V DC. 55 AH/Day @ 6 V DC (other voltages are available)	$533 & up $8745

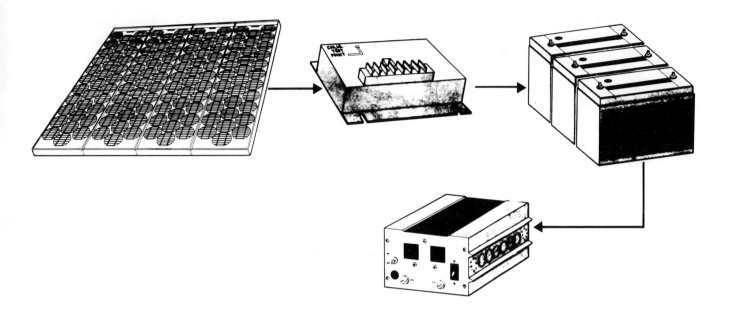

energy efficiency. For example, a typical incandescent AC powered light bulb requires 100 watts of power, whereas a highly efficient special DC powered fluorescent bulb requires about 25 watts to supply the same amount of light.

Although a 12 volt DC system is the least costly and most energy efficient approach to energy independence, many people want to use AC appliances that they already own, such as TV consoles, sophisticated stereo equipment, vacuum cleaners, hair dryers, etc. Since all of the AC power packages use the DC power supplied by the PV panels, it is possible and indeed may be advantageous, to investigate the possibility of installing a hybrid AC/DC system.

Ideally, a hybrid would supply DC power to everything except those appliances that require an AC power supply. This would minimize inverter cost and power losses. Since inverters consume power even if no power is being drawn from them, it is important to place the inverters so they can be easily turned off or use remote or automatic switching mechanisms. If you are considering a hybrid system, make sure that the DC power is available at 12 volts.One of the systems listed in Table 3.7 stores electricity at 12 volts DC and has an inverter for supplying 1000 watts of AC load. This system could easily be used as a hybrid system by connecting a suitable fuse box to the battery.

TABLE 3.7. AC Power Packages

Manufacturer	Application	Description	Rating	Price
ARCO SOLAR	Remote residential-commerical, industrial	1 energy block consists of: 1 KW peak PV modules, 10 Kwhr of batteries, 1 set of mounting structures, 1 ea. std. (NEMA) control housing, 1 charge controller, 2 sets fuses & circuit breakers, 1 set of voltage surge arresters, 1 inverter (chosen from the following):	2Kwhr/day 4Kwhr/day depending on location.	
		25 Series (CAT # 25-A-EBI)	2,500 W continuous	
		50 Series (CAT# 50-A-EBI)	5,000 W continuous	
		100 Series (CAT# 100-A-EBI)	10,000 W continuous	
SOLARWEST Electric Model AC12-12	Remote mobile homes & residences, commercial	12-33 W PV modules (Model ASI 16-2000), 8 batteries (Model Delco 2000), 1 charge controller (Model UCC 12) 1 inverter (Model Best 1000)	90-160 AH/day @12 V DC (available in this rating but fusing must be added) or 120 V AC 60 hz.	$8,400
MODEL AC 16-24		Same as above except, 16 PV modules 12 batteries 1 charge controller (Model UCC 24) 1 inverter (Model Best 2500 W)	24 V DC	$11,000
MODEL AC 24-24		Same as above except: 24 PV Modules 16 batteries	24 V DC not usable for most appliances (use AC only) at rating of not more than 2500 W for 1840-3265 Whrs/day total.	$15,000

3.4 HOME ELECTRIC SYSTEMS

Home electric systems are comprised of the components in the power packages listed above as well as one or more matched applicances. These systems are an especially easy way to get started with PV. Only complete home electric systems are included here. Several firms offer appliances or appliance packages that leave out some key element and are listed in Section 3.5. The starter systems currently available are listed in Table 3.8.

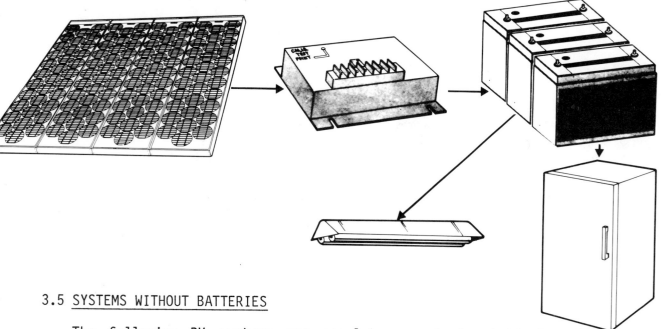

3.5 SYSTEMS WITHOUT BATTERIES

The following PV systems are complete except for batteries that can be purchased separately. Care should be taken in selecting batteries to assure system performance and reliability. Table 3.9 lists some systems without batteries.

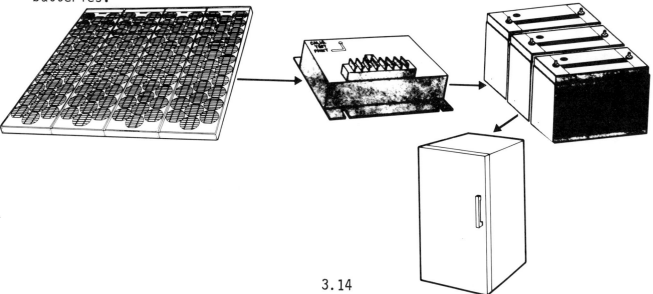

TABLE 3.8. Home Electric Power Packages

Supplier/Mfg. Model & No.	Appliance(s)	Description	Approx. List Price
ARCO SOLAR	Refrigerator & fluorescent lights	1 3.7 cu ft refrigerator (Model WSP 12-6), 2 20 W 1500 lumen fluorescent lights (Model REC 12-6), 4 33 W PV Modules (Model ASI 16-2000), 3 batteries 105 AH (Model Delco 1150), 1 controller, hardware & wiring.	
ARCO SOLAR	Fluorescent lights	Same as above w/o refrigerator, add, 1 40 W 3000 lumen fluorescent light	
Free Energy Systems	Fluorescent lights	1 PV Module (Model 129-SL) 1 battery 28 AH gel cell, 1 8 W 12V fluorescent light.	$499
Motorola	Refrigerator	1-4 cu. ft. refrigerator 4 PV modules 1 support structure 3 Delco 2000 batteries 2 chg. controllers	NA
SOLAREX	Healthcare refrigerator	1 refrigerator/ freezer w/controls & batteries, 3 PV Modules (Powerline)	$5800
Solarwest Electric	Fluorescent light	1 15 W 870 lm fluorescent light (Model REC 115) 1 35 W PV module (Model AST 16-2000) 1 battery 105 AM (Model Delco 2000) 1 controller	$924.50
	Fluorescent light	4 15 W 1050 lm fluorescent lights (Model REC 110) 3 33 W PV modules (Model ASI 16-2000) 3 batteries 105 Ah (Model Delco 2000) 1 controller	

3.15

TABLE 3.9. Partial PV Systems

Supplier/ Mfg.	Appliance	Description	Parts Missing	Price
Jim Cullen Enterprises	Fluorescent & incandescent lighting	1 33 W PV module 1 22 W fluorescent lamp (table) conversion kit. 1 40 W Double bullet light incandescent	Batteries	$500.00
MOTOROLA (MSP 20001)	Fluorescent lights	1 20 W peak PV Module (Model MSP 23A20), 1 control module, 1 20 W 810 lumens fluorescent light, 1 Wiring harness, 1 module support hardware kit.	Batteries 60-80 AH capacity	
MOTOROLA (MSPT20R20)	Television	1 20 W peak PV module (Model MSP23A20), 1 control module, 1 wiring harness 1 module support hardware kit	Batteries 60-80 AH, capacity, TV set 12 V DC	
Solarwest Electric	Fluorescent light	1 33 W PV module 1 charge controller 115 W 870 lumens fluorescent light	Battery 105 AH	$774.50
Standard Collectors Inc.	Refrigerator	3 33 W PV modules, 1 charge controller, 1 refrigerator 5.1 to 7.1 cu ft.	Batteries 85 AH capacity	$2300 with choice of refrigerator.
Western Solar Refrigeration	Fluorescent lights	1 40 W pear PV array, 1 20 W fluorescent light	Batteries.	$975.00
Western Solar Refrigeration	Refrigerator	Solar Modules 4 cu.ft. refrigerators Mtg. & Wiring & Controls Different number of panels for different climates.	Batteries	$4,000 to $6,500
	Fluorescent lights	3 33 W PV Modules 1 charge controller 4 15 W 1050 lumen fluorescent lights	Batteries 105 AH	$1910.00

3.6 SPECIALIZED PV PRODUCTS

Table 3.10 lists some specialized off-the-shelf products that use PV power sources. The list includes several novelty items, educational items, and other particularly useful PV applications.

TABLE 3.10. Specialty PV Appliances

Item	Description	Specifications	Mfg. Supplier	Approx. Price
Parmak Electric Fencer	Completely self-contained electric fencer - in one unit inc. gel battery & solar module.	Controls 25 Miles of fence. Operates up to 21 days in total darkness.	Parker-McCrory Mfg. Co.	$75
AM Radios	Operates from dry battery at night from solar cells during day time.	AM band 1.5 V. operating voltage, dry battery used.	Solec Intern.	$40
	Operates from self-contained rechargeable battery charged by self-contained solar PV module.	AM band with 3 V operation from 2 rechargeable AA batteries	Solarex	
Solar Flashlight	Flashlight contains rechargeable batteries & is supplied with a plug in solar PV charger	2.5 AH rechargeable battery	Solarex	$30
Solar Lantern	Completely self-contained Lantern with rechargeable battery		Solarex	$25
Solar Bird & Game Feeders	Automatic Bird & Game Feeders		Braden Wire & Metal Products	$95 to $335
Petrous "Solaris" Calculator PV powered	Operates in most any location where there is normal lighting level has no batteries	Shell Oil Co.	Shell Oil Co.	$29 using Shell Oil credit card
Teal Photon III Solar calculator	No batteries needed	LCD 8 digits memory & SQ RT	Energy House	$29.95

4.0 BUILDING YOUR OWN SYSTEM

The purpose of this chapter is to assist you in estimating your energy needs and to begin the design of a PV energy system to meet these needs. It is possible that after going through this procedure, you will discover that a PV system is not the most suitable power source for your needs. Before making this decision, you should also read Chapter 7 which reports the actual experiences. However, until you have assessed your energy needs and the utility of a PV power supply, such a determination cannot be made. Consequently, we defer the question of the appropriateness of a PV system to Chapter 6.

Many users are happy with PV Electric Systems (see Chapter 7) who selected their system without going through the procedures outlined in this chapter. They started with what they could afford. However, where requirements are demanding and inflexible, Chapter 4.0 must be used only as a preliminary estimating guide. The process here will be useful in helping you understand PV systems.

There are several tasks involved in designing your own PV system from scratch. We recommend a step by step approach consisting of the following six tasks:

Task 1. Evaluation of Energy Needs

Task 2. Select Appropriate Appliances

Task 3. Array Sizing and Selection

Task 4. Battery Sizing and Selection

Task 5. Select Power Conditioning Equipment

Task 6. Design Review

Although the process has several steps, the arithmetic is simple and the procedure should be straightforward for most readers. The best approach is to generate a rough estimate the first time through, and then refine the calculations on a second round.

Most PV manufacturers supply design guides for the asking that are tailored to their products or, alternatively, will assist serious inquirers in designing their systems with the use of sophisticated computer techniques. This service is extremely valuable and should be used after you have made an estimate using the process outlined here. The guidance offered here will allow you to understand the design process, will help you assess the appropriateness of a PV system, and can enhance your subsequent communications with suppliers.

In this discussion we do not emphasize applications such as warning beacons, navigation aids, etc. It's simply outside the scope of this presentation. There are companies in the business who understand these kinds of problems. They sell complete systems such as warning lights, etc. Two companies with experience in this field are Automatic Power (a Division of Pennwalt Corporation), and Tideland Signal Corporation. They assist customers with "risk and liability management". Properly designed PV systems are reported to be the most reliable power source in these applications, but designing these applications are not for the amateur.

4.1 EVALUATION OF ENERGY NEEDS

The key task involved in designing your own system is deciding how large a system you need or want. In practical terms, this will involve a decision about the size of your initial system, and ultimately how large a system you may need. Although most Americans use large amounts of electricity (particularly by world standards) very few have a notion of how much energy is used for particular applications. This is understandable since electricity, in this country, is about the cheapest in the world and is widely available. Given the rapid escalation of electricity rates, an analysis of your energy needs is useful whether or not you are planning a PV installation.

Another result of cheap electricity has been an apparent lack of concern regarding the energy efficiency of products. Increasing the energy efficiency of most appliances adds to their price; however, the value of the energy saved over the product lifetime usually exceeds this added cost. Consequently, when selecting new appliances for PV systems, energy efficient appliances are highly recommended. Refer to Section 4.2 for more information on appliance energy use.

The goal of this section is to help you estimate daily energy requirements in terms of watt hours/day. This requires an assessment of the following basic factors:

- power requirements of applicances
- anticipated daily period of use for the applicances
- relative need or priority of appliance use
- scheduling of appliance use

Subsequent sections in this chapter will help in accomplishing these tasks, and the example exercise at the end of this chapter will demonstrate how it is accomplished.

As you may recall from Chapter 2, a watt-hour is a unit of energy equivalent to the application of one watt of power for one hour. Since our goal is to determine the average number of watt hours required per day, we need to know the input power requirements (watts) of the appliances to be used and to estimate the average number of hours of daily use. The product of these two factors (derived by multiplication) is the estimate of watt-hours necessary per day.

We suggest that you use the format shown in Table 4.1 to tabulate your energy requirements and categorize the relative level of need. Before filling out this form in earnest we recommend that you finish reading this chapter so that you have some feeling of the system size and cost impacts of appliance selection and scheduling.

The power requirements of most appliances can usually be found in writing on the back of the appliance, in an owners handbook, or from a specifications sheet in terms of watts (see Table 4.2 for some typical power requirements). Some appliances are rated in terms of amps at a specified voltage and in this case the watts can be estimated by multiplying the figures. An important distinction to be made is if the appliance requires AC or DC power. If AC equipment is listed, you must add 15% to the power requirement of the appliance to power the inverter that will be required. Enter the name and power requirements of the appliances you intend to use with a PV system in a form like Table 4.1.

You must also list the AC inverter as an appliance in order to include the "standby power" of this device. It will draw standby power when it is connected to the battery but no appliance is used. The 15% allowance takes care of

TABLE 4.1. Estimate of PV System Size

Appliance	Power Req.	Daily Energy Requirement			Percent w/o Sun	Storage Req.
		Class I	Class II	Class III		

DC Powered

_____

_____

_____

_____

_____

_____

_____

_____

DC   TOTALS
_____

AC Powered

_____

_____

_____

_____

_____

AC TOTALS
_____

****

Power Conditioning

Peak Demands (watts): _____   Inverter Capacity: _____
****
Battery Sizing

Storage Req.: _____ watt-hours   Depth of Charge Limit: _____

Number of Days Storage: _____   Battery Capacity Req.: _____
****
Array Sizing

Average Daily Load: _____ watt-hours   Sun hours per day: _____
Storage Req X .2: _____ watt-hours   Array Size (watts): _____
Adjusted Total: _____ watt-hours   Module Req: _____
****

TABLE 4.2. Power Requirements for Common Appliances

Appliance	Power Requirement
High Efficiency 12V DC Fluorescent	
• 3000 lumens (approx. equal to 150 W Bulb)	36 W
• 2000 lumens (approx. equal to 100 W Bulb)	24 W
• 1500 lumens (approx. equal to 75 W Bulb)	18 W
Medium efficiency 12 V D.C. fluorescent	
• 1700 lumens (slightly more than 75 W Bulb)	26 W
Standard light bulb (at 20 lumens/W)	
• 2000 lumens (example)	100 W
Ovens	
• Microwave oven (120 V AC)	1.35 kW
• Typical electric range (120 V AC)	10 kW
Refrigerators	
• 7.1 cu ft, 12 V DC	60 W
• Standard 120 V AC	500 W
Stereo	
• DC Car Type	15 W
• Home Stereo (AC)	50 W
Evaporative Coolers	
• 5500 CFM	455 W
• 1200 CFM	70 W
Small Household Appliances	
• Coffee perculator 115 V AC	1300 W
• Mr. Coffee 120 V AC	1625 W
• Color TV	60-300 W*
• Toaster	750-1200 W
• Washing Machine	300-600 W
• Fan (small)	25 W
• C.B. Radio 12 V	12 W (transmit)
• Vacuum cleaner AC	700 W
• Hair dryer AC	1400 W

* Varies considerably depending on size and model.

inverter losses while it is supplying a normal load. Standby power can be considerable and can be avoided by using a "demand start" model or by turning the inverter on and off manually (see Section 4.5.2 for a discussion).

The next step requires you to classify the power requirements and estimate the average daily hours of need. You can define the classes of power requirements however you wish, as long as you recognize that all uses are not equally important. The classifications we suggest are as follows:

Class I: Essential Uses not easily scheduled such as refrigeration
Class II: Priority Uses such as exhaust fans and appliances
Class III: Optional Uses such as power tools

The procedure is to estimate the hours of appliance use in each of the classes, multiply by the power requirement and enter the figure under the appropriate use column.

In some instances entries will be more involved than it at first seems. Although refrigerators are used 24 hours/day, they typically draw current for only 6 to 12 hours/day depending upon the refrigerator and the usage. The energy efficiency labels required on new appliances by the Federal Trade Commission may be of help to complete this table. To assist you in estimating the energy needs of appliances or to see how yours compares with energy efficient ones, we refer you to Section 4.2 on appliance selection. Do not hesitate to contact appliance manufacturers before purchase.

There are some appliances that draw so little current that you don't need to worry about them or their period of use is so short as to be negligible in terms of average power consumption. Electric clocks and garbage disposals fall in these categories, respectively. However, it is wise to list them because you may need to account for them when sizing power conditioning equipment.

The timing of appliance use is important for it affects the size of battery storage and the capacity requirements of the power system. For each of the items in your appliance list, estimate what percentage of time they will be used when the sun isn't shining (Table 4.3). You must use your discretion to determine what classes of use you include to calculate storage requirements, but we suggest the sum of classes I and II. Multiplying this percentage times

TABLE 4.3. Example PV System Entry

Appliance	Power Req.	Daily Energy Requirement			Percent w/o Sun	Storage Req.
		Class I	Class II	Class III		
AC Powered						
B&W Television	50*	120	120		50	120

* Fifteen (15) percent is added for inverter power for AC appliances.

the daily use for each appliance and summing provides a baseline estimate for battery storage capacity. The smaller the number, the smaller the battery requirements and cost. Thus, some lifestyle changes that permit daytime use of the system may reduce overall system cost. This effect is discussed in more detail in Section 4.3.

For example, if you are considering the use of a 50 W AC B&W television set, and you consider it essential to watch 2 hours of soap operas per day, and an additional 2 hours in the evening, your entry would appear as in Table 4.3.

Appliance scheduling will also affect the peak demands on the system and, therefore, its capacity rating. The peak demand on the system will occur when the maximum number of watts are drawn at one time. To determine this figure it is necessary to anticipate which combinations of simultaneous appliance use are necessary or desired. Typically, the lower the capacity rating, the lower the system costs. An additional factor that cannot be overlooked is that many AC appliances have higher power requirements when being started than during normal operation. If AC is used, it is important to consider the "surge" power requirements when determining system capacity. More information on this issue is given in Section 4.5.

If you've followed our suggested method, you have no doubt discovered that the arithmetic is much easier than deciding what to include. The technical aspects may be a bit confusing at first but really aren't hard. Making this table is something you must do yourself since each item is likely to be different or unique (some people like TV, others don't, etc.). It is advisable to be realistic, <u>not</u> idealistic, to be happy with the results later on. When you have completed this table, you will be more able to make several important decisions: a) Can you get along with DC alone?, b) If AC is required, can small relatively inexpensive systems be used; for instance, to power a shaver or a radio? c) Do the AC needs outweigh the DC needs so much that a hybrid system will simply be a nuisance?, and d) Can you start with a small system and work up in size over several years taking advantage of the tax breaks to their fullest?

4.2 <u>APPLIANCE SELECTION</u>

The selection of appliances to be powered by PV energy systems may entice you to exercise greater scrutiny than you're accustomed to. The task is not complex or confusing, but does require some additional investigation into the energy efficiency and power demands of the equipment. The information in this section should help you decide to become an informed "comparison shopper." In cases where you may not have the appliances you expect to use or you realize that what you do have requires an inordinate amount of power, be sure to consider the following guidelines in selecting new appliances.

There are four major factors for you to consider in addition to the others you commonly employ. These are:

- whether or not the appliance requires AC or DC current
- the relative energy efficiency of the unit
- the power demands of the unit when it is running
- the percent of time it is actually running.

Clearly not all appliances are alike, even though they may appear identical and offer the same features. What you don't see is the amount of energy required to operate the equipment unless it is labeled with an appliance energy efficiency label. Often you can compare the efficiency of equipment if you look in the owner's manual or on the back of the appliance for the power ratings.

First off, let's discuss the advantages of using DC powered appliances. These typically require a 12 V power supply or run off of batteries of various voltages. For application with PV power supplies (which by nature produce DC

4.8

current), DC appliances may be preferred. In order to run an AC appliance from a PV power supply, an "inverter" must be employed that adds to the expense of the system and reduces the system efficiency. Some DC appliances may not be very efficient or durable, but the quantity and quality are improving rapidly. Some appliances are currently available only with AC power requirements and they must either be foregone or used in conjuction with an inverter. More information on inverters is presented in the next section.

Products for use with a DC power supply are becoming available in growing numbers, primarily due to demand for them by boat or recreational vehicle owners. Most cities now have distributors for such equipment which often have stock on display. We recommend that you look in the yellow pages under Recreational Vehicles to locate distributors in your area. Many have catalogs free for the asking that describe the variety of DC appliances available.

The energy efficiency of AC and DC appliances vary widely. When comparing efficiency you need to determine the energy consumed to accomplish a certain task. Often this is made difficult by differences in the characteristics of the equipment, but if you can express energy use in common units the task is simplified. As mentioned previously, the energy efficiency labels required by the Federal Trade Commision on certain new appliances are invaluable indicators of the relative level of efficiency.

The tables that follow list a number of appliances specifically designed for 12 V DC application. They are generally quite energy efficient. We've deliberately left out 12 V light bulbs that are available from auto and hardware stores although they make perfect sense for intermittent short duty use. We've also not listed 115 V AC energy efficient appliances beginning to appear in most product lines in ordinary stoves. These appliances are typically much better than the older designs they replace but are not generally as well suited to application with small PV power supply systems as the DC appliances listed herein.

It is usually to your advantage to purchase the most efficient appliances available since they permit the power supply system costs to be minimized. Clearly a balance exists between added cost for the efficient appliance and the consequential savings in power system cost. In most all instances you will probably find after going through the sizing calculation that purchase of the most efficient appliance results in the lowest total cost. An illustrative example at the end of this section will bear this out.

Although light fixtures are commonly sold based upon the power (watts) they consume, buyers are really interested in the amount of light (lumens) produced. A common home light bulb (incandescent type) will consume 100 W to provide 2000 lm. This means you get 20 lm/W. A common fluorescent fixture rated at 20 W provides approximately 750 lm of light or about 40 lm/W. The common 120 V AC fluorescent fixture is about twice as efficient as the incandescent bulb and furthermore, should outlast it. Very efficient DC fluorescent light fixtures, available from many RV equipment centers, can be twice as efficient and deliver about 80 lm/W. Those who are discomforted by the "flicker" of fluorescent lights will be glad to know that the high efficiency DC fixtures operate at a "flicker rate" many times as high as conventional bulbs which should be beyond human detection. Producers or distributors of some DC lighting fixtures claiming to produce in excess of 60 lm/W are listed in Table 4.4.

One of the lamps shown in Table 4.4 is a conversion for table lamps and is especially aesthetic in that it produces a warm tone in the room and gives light for working or reading.

Refrigerators are also available with widely differing efficiencies and power requirements. Some models have the capability of operating off of either DC or AC current as well as LPG. Although more expensive than the AC models currently available, they do not need inverters, therefore the DC models maybe more cost effective. In addition the models that operate off of LPG also ensure that refrigerator temperatures will be maintained regardless of the availability of sunshine or the status of the batteries, allowing them to be shifted from class I to class II or III status (in filling out Table 4.1). Some of the DC refrigerators available are listed in Table 4.5. Additional refrigerators were included in Tables 3.8 and 3.9. Be sure to examine the peak power demands (while the compressor is running) as well as the normal amount of time the compressor must run to keep the refrigerator cold.

TABLE 4.4. Efficient DC Light Fixtures

Item	Description	Rating	Supplier	Approximate Price
Fluorescent Lighting				
20 W Fixture	25" unshaded tube and fixture, 83 lm/W	18 W 1500 lm	ARCO Solar	
40 W Fixture	49" unshaded tube and fixture, 83 lm/W	36 W 3000 lm	ARCO Solar	
Rev LII	24.4"x4.2"x4.5", 61 lm/W	19 W 1175 lm	IOTA	$45
22 W	Round 9 1/2" Dia x 6" deep table lamp conversion	15 W 1050 lm	Cullen Enterprises	$42
Rec 110	Round 9.5"Dia x 1.5" Deep Approx. 70 lm/W	15 W 1050 lm	Solarwest Electric	$40
Rec 116	18" Long x 5.5" Wide x 1.4" Deep, 72 lm/W	24 W 1740 lm	Solarwest Electric	$30.50
Rec 147	42" Long x 2.3" Wide x 1.3" Deep, 91 lm/W	24 W 2200 lm	Solarwest Electric	$30
Rec 151	24" Long x 2.75" Wide x 2.4" Deep, 70 lm/W	18 W 1260 lm	Solarwest Electric	$50
Rec 153	48" Long x 2.75" Wide x 2.4" Deep, 87 lm/W	36 W 3150 lm	Solarwest Electric	$65
Fluorescent Thinlite				
#116	18" Long x 5.5" Wide x 1-3/8" Deep, 68 lm/W	26 W 1760 lm	Free Energy, Energy House	$23.33
#126	21" Long x 5.5" Wide x 1" Deep, 66 lm/W	26 W 1720 lm	Free Energy, Energy House	

TABLE 4.5. DC Powered Refrigerators

Company	Model	Size (ft^3)	Top (T) or Front (F) Opening	Power Consumption (Watts)	Energy* Consumption/Day	Price $/Unit	Comments
Norcold	DC 254	4.2 (exterior Vol.)	F	54	650	$315	Most of Norcold's wide range of refrigerators run on AC or DC. Here we've only included their purely DC ref.
Norcold	DC 230	3.3 (exterior Vol.)	F	54	650	$285	
Western Solar Refrigeration	WSR 12-1	4.1	T	72	250-288	$1300	Refrigerator/freezer
Solarwest Electric & ARCO Solar	ARCO Solar WSP-12-6	3.7	T	66	264	$1500	
	Instamatic IE-12V	2.8	F	36	432	$350	
Standard Solar Collectors	J1080-2M	5.2	F	60	172-690	$550-600	
	J1080-7	5.7	F	60	172-690	$560-610	
	J1080-5	5.25	T	60	172-690	$575-620	
	J1080	7.1	T	60	172-690	NA	
Jim Cullen Enterprises, Inc.	Cold Star Cs-4	3.9	F	90	2160	$475	3 way system 110 V AC, 12 V DC, LP Gas. Carries a wide range of refrigerators.

* Based on published data, different ambient conditions may apply.

An illustration of the impact of using a conservation approach is provided by the following example where two systems are compared. The example involves: both a) the use of energy efficient products and b) life style adjustments to minimize energy needs. It is assumed that space and water heating, cooking, and air conditioning are provided by nonelectric methods. The remaining electrical loads are estimated in Table 4.6 for standard and energy conserving appliance use.

It is apparent that adjustments in lifestyle and use of efficient appliances can reduce the energy requirments dramatically allowing one to begin using PV systems in a modest way. The difference is even more impressive when the resultant costs of a PV energy system to fulfill the needs is compared.

In this example the AC power package, AC 12-12, listed in Table 3.7 would handle the 609 Whr with a 50% reserve margin at a system cost of about

TABLE 4.6. Comparison of Standard and Energy Conserving Appliance Use

Lighting	Daily Use (Hrs)	AC Standard System WHrs/Day	Energy Conserving System 12 V DC (WHrs/Day)
5-15 W Fluorescent	3		225
5-75 W Bulbs	3	1125	
Refrigerator 66 W Cooling Unit	4**		264
Refrigerator Standard 500 W Unit	10**	5000	
TV 12" B&W 15 W	4		60
TV 17" Color 50 W*	4	200	
Car Stereo 15 W	4		60
Home Stereo 50 W	4	200	
Dishwasher 1200 W	1	1200	
Handwash Dishes	1		0
TOTAL		7725 Whrs/day	609 Whrs/day

* More energy efficient than most color TVs.
** Based on an approximate duty-cycle of the compressor.

$8,000. About 10,000 Wh/day would be required to handle the standard system and allow for some reserve. This would require modification of the 50 A-EB1 for an additional three energy blocks. The total system cost would be about $45,000 at the present time. Tax credits that can significantly reduce system costs are discussed in Chapter 6.0.

In addition to considering the amount of energy an appliance requires you should also consider the amount of capacity needed. Some appliances require large bursts of power to get them started (AC induction motors for example), whereas others demand relatively small amounts. The lower the capacity needed, usually the less expensive the system is to supply it. When comparing two appliances with otherwise similar features, choose the one with the lowest amperage requirement.

4.3 ARRAY SIZING AND SELECTION

The sizing of a PV array will depend upon the average daily load on the system, the amount of system inefficiency losses, the average amount of sunlight available, and the back-up power supply available. Very sophisticated computer modeling techniques have been developed to calculate array and storage system size, and indeed form the basis for the manual technique described here. Some manufacturers and installers will use these computer tools to assist you in refining your design.

The first step is to determine a grand total for your energy needs from Table 4.1 by adding the AC and DC power requirements. Next add 20% of the storage requirement total to account for energy losses in the battery charge/discharge cycle. Since you've already accounted for the inverter losses, if any, this figure represents the adjusted average daily energy your PV array must supply if no back up is to be used.

The next step is to consult Figures 4.1 and 4.2 to determine the effective (or equivalent) sun hours available in your climatic area. If the system is designed for year round use, and your wintertime loads must be met with the PV array alone, the array should probably be sized according to Figure 4.2 that provides an estimate of the lowest average seasonal levels of sunlight. To calculate array capacity divide the adjusted average daily energy requirement calculated above by the sun hours. This is your array wattage requirement.

$$\frac{\text{Total Average Daily Energy Requirements}}{\text{Number of Sun Hours}} = \text{Array Wattage Requirement}$$

One common myth is that the sun does not shine in Seattle in the winter. It is true that the sunlight is filtered through a dense cloud layer for a number of days per year, but sufficient sunlight is available to produce PV electricity. Only above the Arctic Circle does the sun actually go away in the

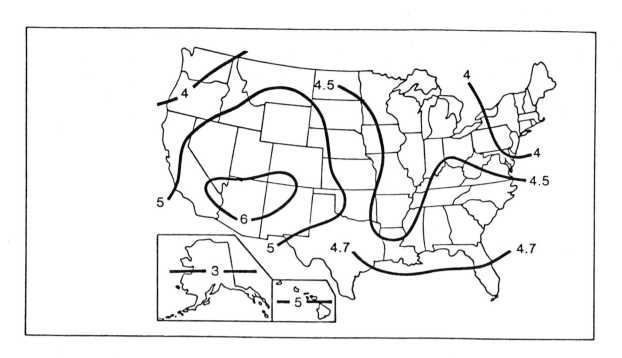

FIGURE 4.1. Peak Sun Hours Per Day - Yearly Average

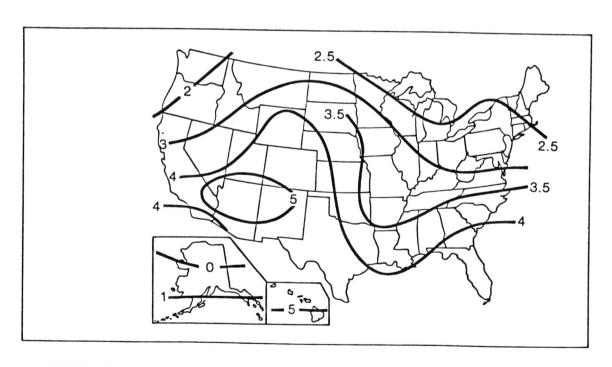

FIGURE 4.2. Peak Sun Hours Per Day - 4 Week Average 12/7-1/4

Courtesy of Solarex Corporation.

winter. No alternative to a backup charger or large battery banks exists for wintertime use of PV systems above the Arctic Circle. A high quality backup generator, 4.5 CM 21 RV, made by Kohler produces 4,500 W and has numerous "quieting" features selling for about $1,800.

To determine the number of modules required divide the array wattage by the wattage of the module. The resultant figure should then rounded to the next whole number. This is your estimated module requirement.

$$\frac{\text{Array Wattage Requirement}}{\text{Module Wattage}} = \text{Number of Modules Required}$$

Check module requirements with the manufacturer or distributor, or use the NASA Report listed in the reference section.

Table 4.7 is a list of firms who either manufacture or distribute PV modules in the U.S., with an example of the type of PV module available. Most companies manufacture a variety of module sizes, however, space limitations prevent us from presenting all of them.

The PV cells are a higher cost portion of modules and arrays as well as the entire PV system (excluding storage). Therefore, it will pay to use ingenuity to derive as much power as possible from each PV cell to reduce the number of cells needed. There are two things that can be done to increase the output of a given cell: a) point it directly at the sun, and b) use a lens or mirror to reflect more light onto the cell.

Devices to direct the array (containing the cells) directly at the sun can effectively lengthen the day so that nearly full output of the array can be obtained from sunup to sundown. This can approximately double the output of energy from a cell under good conditions compared to the usual approach of leaving the array fixed in some compromise position. When lenses or mirrors are used to concentrate the suns rays, the amount of electricity from a cell can be increased nearly directly with the increase in light intensity. Special provisions must be made to carry away the heat and conduct the greater amount of electricity being produced when high concentrations of sunlight are used.

Concentrator methods are under development to reduce costs. When using concentrating collectors, it becomes essential to also track the sun (keep the

TABLE 4.7. Examples of Off-the-Shelf PV Modules

Company	Module Name	Peak Power (Watts)	Voltage (Volts)	Module Size	Application	Approx. List Price
ARCO Solar	ASI 16-2000	35	20	47.9"x11.9"x1.5"	General	$450/Module
ALPHA Energy Systems					Module Distributor/Dealer	
Applied Solar Energy	60-3062	80	32	1.54"x27.38"x47.24	General	
Automatic Power	SP620	4.3	7.2	NA		NA
Jim Cullen Enterprises					Module Distributor/Dealer	$450/module
Energy House					Module Distributor/Dealer	
Free Energy Systems	129 SQ	10	18	15.25"x15.25x0.5"	Marine Applications	$399/Module
Glass Energy Electronics					Module Distributor/Dealer	
International Rectifier					Custom Modules	
Mobil/Tyco	Ra40	40	10-21	42"x17"x2"	General	$1000/Module
Photon Power	Photon Power	80-100		4.5'x8.5'	An 8-Panel Module	$5-$12/Wp
Photowatt	MB3310	18.6	20.5	11.3"x36"x.875"	General	$304/Module
Silicon Sensors	SPP3300-12	33		30"x24"x2"	General	$500/Module
Silonex	NSL-5925-32		19.5	47x15x1.5	General	$670/Module
Solar Power Corp.	G12-361	33.5	16.5	44"x17"x2.25"	General	
Solar Usage Now					Module Distributor/Dealer	
Solarex	PL 100	60	6.6	26"x 51"	General	$1540/Module
Solavolt	MSP43E40	40	19.5	14"x48"x1.5"	General	$650/Module
Solec Intern.	S-4134	32	20.5	37"x16"x1"	General	$600/Module
Solenergy Corp.	SG1294-G	33	12 (NOM)	23.6"x30.2"x1.75"	General	
Solarwest	ASI 16-2000	35	20	48"x12"x1.5"	Module Distributor/Dealer	$450/Module
Sollos	P264	10-12	12.75	18"x18"x.5"	Irrigation	NA
SPIRE Corp.	Hi Performance	60	22	47"x16"x2.4"	General	
Tideland Signal	MG-300/12	3.8	13.4	10"x10"	Navigational Aids	$242/Module
United Energy Corp.	1212	20.3	14.5	25"x15"x3/8"	General	$400/Module

lens between the sun and the cell). A passive solar tracking device is available from Zomeworks. Several manufacturers have supplied information about concentrators summarized in Table 4.8.

Standard modules gather sunlight on hazy or cloudy days. It is quite surprising to observe a 2 amp panel producing a little less than a quarter ampere in Seattle, Washington, with complete cloud cover. Ordinarily concentration systems do not work as well under cloudy or hazy conditions. Most concentrator systems are presently being recommended by suppliers only for larger installations.

There are several things to consider in purchasing a concentrator array:

- Reliability of tracking apparatus
- Maintenance and service requirements of concentrators and tracking apparatus
- Performance in the specific climate conditions encountered

TABLE 4.8. Concentrators and Arrays

Model/ Supplier	Description	Approximate Concentration Ratio
Automatic Power "Solar Power Concentrators"	Removeable aluminum mirrors bolt to SP 320 or SP 620 automatic power modules.	About 1.5 without tracking or adjustment
Solarex C 250 Array	6 panels produce 250 W peak output flat mirrors attach above & below modules.	About 1.5 without tracking or adjustment
Acurex 3001-P	Parabolic through actively cooled concentrating array with single axis tracking, can thus supply hot water in addition to electricity	40
AAI PV/thermal Linear Concentrating	Linear mirrors focus sun on a line of PV concentrator cells providing both electricity and hot water at 190^{0}F single axis tracking	40
Acurex Solar PV Concentrator	Tracking array (2 axis) of sealed beam headlight type PV concentrators	100

- Appearance factors (a pedestal tracking array looks unusual in some neighborhoods)
- Any unusual environmental conditions.

If you have any questions regarding the performance of your module(s), you can send them to DSET Laboratories, Inc. for testing and other assistance.

4.4 BATTERY SIZING AND SELECTION

This section provides a method by which you can estimate the size of battery storage needed and provides some guidance in the selection of batteries. The reader is cautioned that battery technology is quite complex, and the variety of batteries available is varied. Consequently, investing time in the study and selection of batteries is very worthwhile.

We cannot possibly present and explain in this directory all battery issues and products available, but we will attempt to raise issues of paramount importance and offer some guidance in the design and selection of electrical storage batteries. If you've completed Table 4.1, you are well on your way to determining the battery storage requirements.

4.4.1 Sizing Battery Storage

If you estimate the percentage of time that appliances will be used during non-sunlight hours and multiply this figure times the average watt-hours used per day, you have determined the amount of energy that typically needs to be removed from battery storage each night for that appliance. By adding this figure for all the appliances, we derive the total amount of energy needed to be removed from the battery storage system on a typical daily cycle.

The life and, consequently, cost effectiveness of batteries is based upon: a) the number of times the battery is discharged or "cycled", b) the percentage of the total battery charge (depth of discharge) that is removed during the cycle and, c) the quality of maintenance. Each type of battery responds differently. So the question that needs to be answered is "What depth of discharge is best to maximize battery life and minimize storage size?" Given the variety of batteries available with different characteristics and the widely divergent loads placed upon them, a simple universal answer is not available.   The question is best resolved by discussions with battery manufacturers or suppliers.

In "typical" PV applications, ideal practice is to specify good quality batteries experiencing a 10% average depth of discharge that will last approximately 5 years.   Using fewer batteries and increasing the depth of discharge to provide the same amount of energy daily may shorten battery life and reliability but may still be cost effective.   If you use batteries and further reduce the average discharge depth, the cost of the storage system increases to a point that may not be recovered by anticipated extensions in battery life.

In actual practice, deep cycle batteries are being discharged more than 10% on a daily basis with fairly good life that saves considerabley on system first costs.

The actual depth of discharge in a PV system can vary from 0 to as much as 90% of capacity if the battery is designed for this, since an occasional deep discharge isn't too harmful.   However, one is faced with the inescapable trade-off between battery life and depth of discharge.   Leaving batteries in the discharged state for extended periods causes premature failure.   Some batteries can stand more of this type of use than others.

One way to ensure that batteries are not discharged excessively so that life can be maximized is to install an alarm system or automatic shut off mechanism (as described in Section 4.5.3) to operate when the battery is at a specified depth of discharge. A backup system such as a gasoline generator set can then be employed to return the batteries to full charge while providing for the appliance electrical needs.   If no back up is employed, the designer has two options: 1) discharge the batteries more deeply (and pay for it in reduced life expectancy), 2) install additional battery capacity or, 3) wait for more sunshine.

The second approach may be perfectly acceptable because of the modular nature of battery systems. If it is found that more power is frequently needed than can be provided by a reasonable depth of discharge, additional batteries can be added.   Of course, if the charging system (say the PV array) is inadequate to return the batteries to full charge in a typical daily cycle, then it must also be enlarged.

To determine initial system sizing, make an estimate of daily storage requirements from Table 4.1 and multiply by 10 to get an estimate of the total battery system capacity, i.e., if you have a backup generator. If no backup is available, multiply anticipated essential energy needs by a factor of 15 to 25 to ensure adequate battery life. If the sun is frequently obscured and a backup battery charger is not available to provide for these cloudy periods, the array and battery system size must be increased.

One final check to make after sizing a battery system is to see if the maximum power output (amperes) of the PV array is less than 1/20th of the total amp-hour capacity of the battery system. This is because the maximum constant charge rate for conventional lead acid batteries should not exceed about 5% of the battery capacity.

The following section describes the key factors to keep in mind in selecting and using a battery system.

4.4.2 Factors to Consider in Selecting and Using a Battery System

Automotive-type batteries are frequently not suitable for PV use. PV battery systems need:

- **Deep Discharge Capability.** Automotive-type batteries are seriously damaged if fully discharged and left for any period of time. They are not designed for PV use. In a PV system the battery may be fully discharged occasionally.

- **Adequate Life.** The life will be related to the number of charge/discharge cycles and to the depth of the discharge/charge cycle.

- **High Charge/Discharge Cycle Efficiency.** The efficiency of the battery and charge controller together should be about 80% or more for normal operation. If this isn't true, a larger PV array would be needed to make up the difference. A good PV battery will internally consume only 2% to 4% of its full charge per month. Automotive-type batteries may self-discharge as much as 30% of their full charge in a month without any outside power load.

- <u>Maintenance</u>. Maintenance free batteries are available for PV use, and may be important for use by those who don't intend to become "knowledgeable" about storage batteries. Other types of batteries absolutely require some maintenance (water level adjustment etc.) to prevent premature failure.

- <u>Safety</u>. Batteries are safely used by the millions in automobiles. They are inherently safe when used properly, but they typically contain strong acids, and they can generate hydrogen gas which is highly explosive. Large quantities of batteries should be isolated from living quarters in well vented enclosures. Suppliers should be consulted about housing and protecting batteries. Ordinary batteries can be fitted with "hydrocaps" which help them behave like maintenance free batteries. They are available from Hydrocap Corporation.

- <u>Freeze Protection and Shading</u>. In northern climates batteries need freeze protection. Batteries fully charged are quite resistant to freezing although they may have drastically reduced capacity. Your car may not start well on cold mornings because of this problem. Specially designed batteries can be obtained so that they won't freeze even if discharged; however, they still won't put out full power when cold. It may well pay to build a separate underground protective enclosure in northern climates where freezing and cold weather are common. In warmer climates an underground enclosure can prevent excessive over-heating and help prolong satisfactory battery operation with reduced maintenance.

- <u>Convenience</u>. In some uses, gel electrolyte batteries are convenient since they can be over-turned without spilling corrosive acid.

When buying a battery system, there are several factors to consider and you should be aware of the factors listed below when examining the various battery systems offered.

- <u>Cost and Life</u>. What are the tradeoffs between initial costs and life expectancy?

- <u>Storage Capacity</u>. How many days can you operate "without sunshine"?

- <u>Efficiency</u>. Does the battery selection affect the size of PV array needed?

- <u>Ruggedness</u>. Can the battery withstand the conditions of use; i.e., the amount of discharge it may encounter, etc.? Is it physically designed to cope?

- <u>Safety and Reliability</u>. What costs, if any, are required to make the system safe and reliable? Do you need a special building, etc.?

- <u>Maintenance</u>. How much maintenance is required?

The battery field is very dynamic and we cannot possibly list all suitable batteries. Several firms that specifically contacted us and asked to be mentioned as suppliers of P.V. batteries are included in Table 4.9.

<u>TABLE 4.9</u>. Example Suppliers of PV Batteries

Supplier	Description & Comments	Rating	Approx. Price
C&D Battery Co.	Type "Q,P" stationary very deep discharge batteries (extra heavy duty)	96-936 Ah	$165 to $522
C&D Battery Co.	Type DCP, KCP, LCP medium discharge long life batteries	42-2880 Ah	$143 to $1015
Cullen Enterprises	8DVDH medium discharge & 8DVXXH deep cycle, long life	220-300 Ah	$190 to $260
Solarwest Electric and ARCO Solar	Delco Type 1 batteries maintenance-free construction in "automotive style" case	105 Ah 12 V DC	$165
Exide/Energy House	PHv DE 30 6 V deep cycle battery in "automotive style" case	6 V, 185 Ah (use 2 for 12 V, @ 185 Ah)	2 for about $200
Free Energy Systems, Inc.	Globe Union, Gel Cell deep discharge type, sealed, "long life"	28 Ah 12 V	
Gates	0800-0016 sealed 12 V lead acid batteries	5.2 Ah @ 12 V DC	NA
Surrette	301 M marine battery	6 V 230 Ah (2 for 12 V, 230 Ah)	2 for about $200

The solar module supplier or dealer may make recommendations of additional suppliers of batteries. The book, How to Design an Independent Power System, by T. D. Paul has many practical suggestions about batteries (see reading list).

4.5 SELECTING POWER CONDITIONING EQUIPMENT

This section discusses the various power conditioning equipment that is necessary and available to ensure stable, safe, and efficient operation. The equipment falls into three categories; power controllers, inverters, and battery monitors.

4.5.1 Power Controllers

The use of PV panels to charge batteries wouldn't require controls or electronic devices if the panels could be disconnected at night and in the afternoon. This will keep the modules from using power out of the battery at night and prevent the batteries from overcharging in the afternoon. It is important to reconnect the array to the battery when the battery gets "low". In some applications such as warning beacons, batteries and PV arrays can be matched so well that the batteries will never overcharge.

The ARCO Company has designed a little black box to do just exactly that task for you. It has a switch (actually a relay) that is controlled by an electronic circuit called a universal charge controller (UCC). Most UCCs are available to operate on voltages of 12-48 volts nominal system voltage. It can even be ordered with a special probe to sense the temperature of the batteries, and to adjust its operation to the exact voltage the battery should have when fully charged at that temperature. It can handle up to 25 A of current from panels at 12 V DC and up to 1 kW of panels at 48 V DC nominal voltage. The UCC unit sells for about $325 from Solarwest Electric.

However, there are disadvantages to this device -- it does have mechanical contacts whereas approaches subsequently described are solid-state and eliminate this problem (but have other disadvantages). It uses about 70 mA at 12 V DC during charging and uses 15 mA at 12 V DC the rest of the time.

The most widely used approach for small systems is to omit the charge controller but to place a "diode" (or one way valve) in series with the array to keep the current from flowing back out of the battery and through the array at night. This diode is usually called a blocking diode and can be ordered right in the module. Small systems seldom overcharge the battery and, if battery water levels are maintained, it is not a serious problem. Diodes are not needed and shouldn't be used with the UCC described above because they would consume power during the charging process.

If a system is to be used under changing load conditions some sort of charge control is needed in addition to the "blocking diode." A commonly used approach is the "shunt regulator". The device turns on some extra load when the battery is fully charged and the voltage gets too high. It "shunts" or bypasses the current away from the battery until the battery gets low and needs to be charged again. Some people will be able to remember when automobiles had no regulation of battery voltage and on long trips it was advisable to turn on the headlights to keep from boiling away the battery water. The shunt regulator does a similar job using a solid-state switch to connect a resistor that consumes the excess power.

The disadvantage of the shunt regulator is that it generates a lot of heat that must be dissipated. For this reason, they are fairly large. A line of shunt regulators available from Solar Power Corporation is shown in Table 4.10. They are less expensive than the UCC for small systems. These

TABLE 4.10. Solar Power Corporation Voltage Regulators

Model Number	Typical Regulating Voltage (Volts, DC)	Current (Amps)
BVR12-0.6	14.7	0.7
BVR6-1.2	7.35	1.4
BVR12-2.0	14.4	2.5
BVR12-2H	14.4	2.5
BVR12-6.0	14.4	7.35
BVR12-12*	14.4	14.4
BVR24-0.7	28.8	0.7
BVR24-2.0	28.8	2.5
BVR24-6.0	28.8	7.5

*Supersedes Model BVR12-10

shunt regulators can be ordered with temperature compensation and high and low voltage alarms. They are also available with special adjustments for operating voltages.

Motorola Corporation has designed a clever shunt voltage regultor system with one master unit to dissipate as much as 100 W that can control as many 80 W slave units as needed to keep the battery bank from overcharging and to accomodate the charging current. The master MSPR 125 LXX is available for about $125 each and the slave unit (KISPR 125LS) costs about $100 each. Complete information can be obtained from Motorola or their dealer in your area. These units do have temperature compensation and contain "blocking diodes". With this system blocking diodes don't have to be ordered with the modules. Regulators can be ordered in sufficient quantity to prevent overloading.

Considerable creativity is emerging in the area of controllers. Jim Cullen Enterprises has several controllers including one that dumps excess power into a resistance coil inserted into a hot water tank.

Arco Solar has a "Battery Protector" load management controller that will switch off low priority loads when the battery is low and reconnect them when the battery is properly charged.

Solarwest Electric and Specialty Concepts have developed charge controllers. "The Sun Charger" by Specialty Concepts is designed to switch loads from the battery to the solar modules, when they are producing enough power, so that the power does not go through the battery.

It is advantageous to discuss the system to be regulated with the regulator supplier to assure satisfactory results. The module/array supplier will usually design and supply a regulator (for a fee) to meet your specific needs. The previous discussion won't really be needed in this case, but it may help you discuss your needs with the supplier. He will want to know the battery type (lead acid, lead calcium, etc.) and will, of course, know the array size you are ordering and be able to supply adequate voltage regulator capacity for excellent system performance.

Power controllers are unnecessary if batteries are not used, and the devices to be connected can tolerate the high voltage generated by an unloaded panel. The so called "sun synchronous" application, with a motor or other load matched up to the panel, will have no difficulty under normal conditions. An unloaded 1/2 hp DC motor may not load a 450 W array enough to keep the voltage in the normal operating range, but shunt-type battery charge controllers can be used for this purpose even though batteries aren't used. We recommend that you discuss these applications with the module supplier.

4.5.2 DC to AC Conversion

The most efficient way to use the power generated by PV panels, in small systems, is to use it in the DC form in which it is generated. For this reason many PV systems have started out as pure 12 V DC systems. Most of the major needs such as pumping, refrigerating, and lighting can be provided by 12 V DC appliances and products. However, sooner or later the user of a pure DC system will want to operate AC appliances. It is quite natural to want to add some of the regular AC appliances over a period of time and a straightforward way to handle this problem is to add devices to change some of the DC to 120 V AC. These devices are called inverters.

A wide variety of inverters are available and we will decribe some representative inverters for major uses. Most PV module suppliers can provide inverters or suggest sources.

Inverters naturally fall into the classes listed below:

- Superlight duty for intermittent use with specific small appliances such as electric razors -- operated from battery systems.
- Medium duty for use with light hand drills, etc. and small appliances that have modest starting surges -- operated from 12 V battery systems.
- Heavy duty inverters for use in starting and operating heavy duty appliances such as large AC refrigerators, air conditioners, table saws, etc. that have a requirement for starting surges about six times the normal full load current -- operated from a DC battery bank.

4.27

- Low cost inverters that connect to the electric power company lines don't use batteries at all and can't operate when the utility power is shut down.

- "High quality*" inverters that can connect to the power line or operate in a "stand alone" mode connected to a battery

The superlight inverter is made to be used right with the appliance such as an electric razor. Since inverters do draw current, they should be shut off when the appliance is not in use. No special house wiring is required and they do the job very well within the limits of their design. Some of these super-light inverters are listed in Table 4.11.

---

* The "high quality" sine wave inverter may be better for operating computers or special equipment. However, heavy duty inverters may be a better choice for starting motors according to published accounts.

TABLE 4.11. Superlight Inverters from Jim Cullen Enterprises

Supplier	Description	Output Wattage (Cont.)	Price
ATR	Type 12U-RSF complete with cables	150	$141
	Type 12-DME with cigar-ette lighter adapter	40	$ 65
	Type 12-RME-1 with cigar-ette lighter adapter	110	$ 99
	Type IL-SPB "Shav Pak" with cigarette lighter adapter	25	$ 53
	Sine wave adapter produces "true" sine wave with above		$ 60
Tripp-Lite	PV 115	100	$ 60
	PV 200	200	$ 75
	PV 500	500	$159

Inverters consume power if left connected even though the appliance is turned off. They may be fairly inefficient, but this effect is small if use is limited to short duration on an occasional basis. Inverters can generate considerable noise, both audible and electrical. The electrical noise may be translated into audible noise ("static") in a radio or high-fi.

The medium duty inverters are available for operating small appliances, electric razors, light power tools, etc. They should be operated below their maximum rating and turned off when not in use. Some inverters are available with a "load demand start" feature that allows use without turning the inverter on or off. This feature should be ordered even though it costs a little extra since it can save your battery supply by eliminating the consumption of standby power. A word of caution is in order here! Demand control of start and stop works well for "normal" loads, but users report that for very light loads (i.e., small radio), some automatic start controls do not start. Manual override switches would appear to be a solution to the problem. The units listed in Table 4.12 don't have enough surge capacity for starting large AC induction motors.

TABLE 4.12. Medium Duty Inverters 12 V DC to 120 V AC

Model # & Supplier	Maximum Watts Continuous Output	Maximum Watts Peak Output	Input Watts No Load	Approx.* Price
PV-250 FC Solarwest Electric	250	275	18	$185
1057, Wilmore	300	NA	NA	$400
PV-500 FC Solarwest Electric	500	550	42	$300
1403, Wilmore	500	1000	NA	$600
PV-1000 FC Solarwest Electric	1000	1100	42	$450
B-5 Dynamote	500	1000	Available with demand control of start & stop.	
B-10 Dynamote	1000	2000		$840
B-18 Dynamote	1800	3500		$990
Best B-12-1000 Solarwest Electric	1000	4000	18 W	$1,199

* Without load demand start and stop (add $100-$200 for this feature).

Heavy duty inverters are required to start large AC induction motors because they require as much as six times the normal running current (i.e. surge) starting period. Table 4.13 supplied by T.D. Paul of Best Energy Systems compares the starting and running watts for typical AC induction motors. He has a number of practical suggestions in his book, How to Design an Independent Power System, for those considering their own systems.

TABLE 4.13. Comparison of Starting and Running Power for AC Induction Motor

Horsepower	Start/Run Watts	Horsepower	Start/Run Watts
1/6	1500/325	1/2	6000/1000
1/4	3000/525	3/4	8000/1400
1/3	4500/725	1	10,000/1800

Table 4.14 lists some representative units that have high surge capacity.

Model & Supplier	Maximum Continuous Load	Surge Load	Battery Voltage Nominal	No Load Input (W)	Universal Tool Rating	AC Induction Motor Rating	Approx. Price
Dynamote: B18-24	1,800	3,600 W	24	Available with load demand control of starting & stopping.	NA	0.5 HP	$ 990
Best/Solar Electric: BR-24-2500	2,500	10,000	24	40**	3 HP	0.5 HP	1,749
Dynamote: B30-24	3,000	6,000	24	Available with load demand control of starting & stopping.	NA	1 HP	1,350
Best/Solar Electric: B-48-5000	5,000	20,000	48	70**	6 HP	2 HP	2,720
Dynamote: MB60-48*	6,000	18,000	48	Available with load demand control of starting & stopping.	NA	2 HP	2,400

*Includes load demand start and stop control and other features to make it suitable for marine use.
**Manufacturer is adding load demand start.

The invertors we have discussed up to this point are all solid-state units and most claim efficiencies in the range of 85% for loads of 25% or more of full continuous load inverter rating. "Wave shape" produced by these inverters is not a "sine wave" (the smooth ideal wave shape for AC). They all have some variation from this ideal. The amount of variation partially explains the differences in price. Best's B-48-5000 and Dymamote's B-60-48 has been specifically recommended by the manufacturer as having the capability to <u>start</u> and <u>operate</u> typical household appliances. If a 5,000 W range inverter must be left on to operate freezers, etc., it will consume up to 1.7 kWh/day just in standby power. If left turned on 24 hrs/day, it would quickly discharge the batteries of a small PV systems even without a load connected. Because of this, some users have two inverters, a small one to operate small appliances and a large one to operate larger equipment on an occasional basis. Therefore, hybrid systems are more practical for small systems that use DC appliances and AC inverters are best for powering equipment unsuitably powered from DC. In large systems with many loads, it is probably simpler to operate the inverter full time and add enough panels to accomodate the loads and standby power of the inverter. All of the units listed so far were designed to operate from storage batteries, and none was intended for connection to a power line to feed excess power back to the utility.

<u>Low cost inverters for connecting utilities</u> are a special type of inverter for feeding power from a DC source into the utility grid. The advantages of this design are: a) low cost, b) its ability to feed power back into a utility system from a PV array without any battery storage, c) the ability to handle wide voltage changes, and d) very high efficiency. The disadvantages are that this design will not operate away from the utility and the "wave shape" of the current produced isn't the same as that produced by the utilities. This could produce special problems. If several inverters were to be operated from the same utility distribution transformer, serious problems might be encountered. On the other hand, your utility may have no problem at all (at least 600 are operating in the U.S. at the present time). In any case don't plan to connect an inverter to the utility AC power system without checking with the utility first. These inverters are called the Gemini

series and are produced by WINDWORKS, Inc. Their approximate prices range is from about \$1/watt in smaller sizes to about \$0.18/W for the 150 kW unit. This is well below the cost for other types of synchronous (or utility connected) inverters. Some of the ratings available and approximate prices are shown in Table 4.15.

There are a number of developments in "high quality"* inverters that will bring additional products into the market where good "wave form" is produced and the inverter is to be operated with or without AC power line inter-connections. The Nova Electric Manufacturing Company and the Topaz, Powerwork Divison make a number of "high quality" sine wave inverters. These are listed in Table 4.16. Abacus Controls makes a line of DC to AC inverters (Table 4.16) to operate with a DC input voltage of 120 V and another Sunverter line operating with 200 V DC (nominal) input (160-240 V DC with or without battery). Part of the Sunverter line is described in Table 4.16. Prices are about \$5/volt ampere where VA is a product of load voltage in volts and load current in amperes at a 250 VA rating and go down to about \$1.70/VA at 10,000 VA rating. The "suninverters" can operate in a stand alone mode or can be supplied to

---

* The "high quality" refers to purity of wave shape which is not needed in many applications.

TABLE 4.15. Ratings Available and Approximate Prices for Gemini Synchronous Inverters

Ratings (kW)	Approximate Price
4	\$ 4,072
8	4,936
10	6,180
15	8,393
20	12,000
40	13,700
50	16,500
100	22,100
150	27,000

TABLE 4.16. High Quality Sine Wave Inverters
120 V AC, 60 Hz output

Model and Supplier	Output Power (VA)	Input Voltage	Approximate List Price$
1260-12 Nova	125	12	485
250 GZ 12/24 60-15 Topaz	250	12/24	1,190
2560-12 Nova	250	12	690
500 GZ 12/24- 60-115 Topaz	500	12/24	1,720
5060-12 Nova	500	12	975
1000 GZ-12/24- 60-115 Topaz	1,000	12/24	2,310
1K60-12 Nova	1,000	12	1,765
443-4-120 Abacus	4,000	120	7,130
5K60-120 Nova	5,000	120	5,070
463-4-120 Abacus	6,000	120	10,380
10K 60-120 Nova	10,000	120	9,145
483-4-120 Abacus	10,000	120	13,820
743-4-200 Abacus	4,000	160-240*	9,360
763-4-200 Abacus	6,000	160-240*	12,240
714-4-200 Abacus	10,000	160-240*	17,050

* These are "sunverters" and operate directly from PV arrays with or without battery banks and can feed power back into a utility grid.

operate connected to a utility power line. The wave form is much closer to the ideal with this type of inverter than with the lower cost version. In spite of this do not plan to connect to the power company without checking with the power   company   first.

American Power Conversion Corporation and Advanced Energy Corporation are reported to be close to issuing catalogs on synchronous and stand alone inverters in the 2 to 8 Kw range.

A special word of caution here, we recommend that you get assistance from experienced designers and installers of these inverters that connect to the power line. There are many potential problems we can't discuss here including safety issues, laws, rules, and regulations that must be followed in these installations. The inverter manufacturers may be willing to assist with these problems.

The problems associated with smaller stand alone inverters are more manageable by the knowledgeable do-it-yourselfer. They are being used by do-it-yourselfers already. This isn't the case for the utility connected inverters right now. As time passes, the problems will be better understood and less skill will be required to install these utility connected systems. In addition to Windworks and Abacus, there are other companies able to design inverters for various applications including Helionetics and Applied Research and Technology Corporation of Utah (ARTU).

In summary, inverters can supply AC from DC sources for most reasonable home and light industry needs. At present PV costs, it is better to use high efficiency DC appliances where possible, especially where the use is fairly continuous. Occasional use appliances can be supplied with AC from very efficient inverters and add much to the enjoyment of the PV power system. A very small inverter (200 W) can supply a sewing machine or portable mixer.

4.5.3 DC to DC Conversion

Selecting system DC voltages presents some challenge. There are advantages and disadvantages to any voltage that may be selected. A choice is available wherever several 12 V DC solar panels and 12 V batteries are used to store the energy. The panels can be connected in parallel to charge the batteries at 12 V or connected in series for a higher voltage: 24, 36, 48, etc.

The 12 V DC system has considerable advantage in reduced shock hazard, and because there are numerous appliances designed for 12 V DC use. Two problems exist with the 12 V system choice. One is that larger wires are required for carrying the lower voltage, and in larger PV systems heavy duty inverters are often selected that must be supplied with a voltage higher than twelve.

If a higher voltage is selected, a DC to DC converter will be needed to produce 12 V for many DC appliances.

A seemingly simple solution to the problem is to hook a resistor in series with the appliances to be powered from the higher voltage source. This approach is not recommended because it can destroy the appliance under certain conditions. Although there are situations where this approach can be used safely, it is especially hazardous for sensitive electronic equipment with a wide variation in current consumption during normal operation. For instance, a C.B. radio may use twice as much current during transmission compared to standby and listen mode. In this case, if you select the dropping resistor for the correct voltage during listen operation, the radio would get only about half voltage during attempts to transmit. Selecting the resistor for transmission mode would greatly overvoltage the appliance during the receiving mode.

DC to DC converters can match the supply voltage to the load voltage. They are available in a wide variety of voltages and power ratings. They do an excellent job of providing DC power for various appliances, but they consume some power during standby mode and should be switched off when the appliance is not needed. During operation they are typically about 80% efficient. Some examples of DC to DC converters are listed in Table 4.17.

Many other converters and power supplies are available from Wilmore Electronics and other sources.

DC to DC converters are a satisfactory solution to the problems of operating DC appliances but they do have three disadvantages in that they: a) lose about 20% of the power they convert while being used, b) use power even during standby mode with no load attached, and c) add more complexity to the system and consequently add to system expense. Despite these disadvantages, they are likely to be used extensively as typical PV systems grow in size and power ratings and high-powered AC inverters demand higher DC supply voltages.

TABLE 4.17. DC to DC Inverters

Item # and Supplier	Nominal Supply Voltage	Nominal Output Voltage	Output Current	Standby Power (No Load)	Approx. Price
8424 Solarwest Electric	24 DC	12 V DC	12 A Peak	NA	$250
Model 1365 Wilmore Electronics	24 DC	12 V DC	10 A Continuous 15A 20% Duty Cycle	4 Watts	$175
Model 1265 Wilmore Electronics	24 or 48 V DC (specify)	12* V DC	30A	5 Watts	$648
Model 12-150 Pacific Energy Systems	16-40 V DC	13.8 V DC	11 A	1-2 Watts	$198

* Voltage is adjustable from 12 V DC to 14 V DC by a screwdriver adjustment on the front panel and output is highly regulated from no load to full load from 20-29 V DC input or 40 to 58 V DC for the higher voltage model.

4.5.4 Battery Monitors

Alarm systems can be obtained to warn of low voltage or high voltage or of a 24 hour period with no power coming from PV modules. The module manufacturers are usually able to provide alarm capability. We are aware of two models available from ARCO Solar for 24 V and 48 V systems. The "AM 24" or "AM 48" alarm device is well worth the $200 to $300 purchase price. Dymamote also lists voltage guards for 12, 24 and 36V systems. They can warn that some loads must be dropped or other action taken.

4.6 DESIGN REVIEW

The purpose of this section is to assist you review your electrical needs and design a PV power system to meet them. First, we will highlight the most important factors from the previous sections, and then present an example calculation to demonstrate the design process. It will be helpful for you to refer to a copy of Table 4.1 as you study this material.

4.6.1 Estimating and Reviewing System Sizing

The first task to be accomplished is to estimate the anticipated energy requirements. This is done by listing the appliances you intend to use in Table 4.1 according to the type of power (AC or DC) required. Enter the power requirements (watts) of the appliances under the second column. Remember to add 15% (multiply requirements by 1.15) of the power requirements of AC appliances listed to account for the power consumption of the required inverter.

Next you must estimate the hours of use for each of these appliances or Class III (optional). Under each of these classes enter the product of the number of hours times the power requirement for each appliance. This should be expressed in terms of watt-hours because you've multiplied the power requirement in watts times the number of hours of average daily use.

The next step is to estimate the storage requirements to size the battery storage system. For each appliance, estimate the percentage of time the appliance will be used when the sun is not shining. By multiplying this percentage by the daily energy requirement (i.e. watt hours) calculated above, you have a preliminary estimate of the amount of energy which needs to be extracted from batteries on an average daily cycle for each individual appliance.

Now total up the columns for the DC appliances and the AC appliances and enter them on the table. The next task requires you to anticipate the maximum power requirement which will occur with your system by adding up the requirements of those appliances which will possibly operate at the same time. The resulting total for the AC appliances only is your inverter capacity requirement which should be entered in the power conditioning section. The peak simultaneous power requirement for both the AC and DC appliances should be inserted in the peak demand blank.

According to our rule of thumb, which is based upon conventional practice, we estimate the battery storage requirement. Enter the total of the storage requirement column (AC plus DC) in the Storage requirement blank in the battery sizing section. Enter 10% for the depth of charge limit and multiply your storage requirement by 10 to determine your battery capacity requirement. If you live in a frequently cloudy area estimate the number of days that you expect cloudy conditions and enter in the number of days of storage blank. Add 10 to this number and multiply by the storage requirement to estimate the battery storage requirement.

Finally, we need to size the PV array required to keep the batteries sufficiently charged. Enter the sum of the DC and AC energy requirements in Class I and II and enter in the Average Daily Load blank under Array Sizing. Multiply the storage requirement calculated above by 0.20 and enter the result below. Sum the average daily load and the 0.2 times storage requirement figure and enter as the Adjusted Total.

The next step is to consult Figures 4.1 and 4.2 to determine the average sun hours in your location. If the PV system must provide all of the Class I and Class II needs at all times you must use the lowest of the sun hour figures. By dividing this figure into the Adjusted Total, you determine the Array size necessary in watts. By dividing this figure by the output watts of the modules selected, and rounding you have an estimate of the number of module panels needed for your installation.

By looking back through the tables on products available it is possible to estimate the cost of your system. Having gone through this exercise, you may wish to reconsider the appliances used or the hours of use to reduce the cost of the PV supply. An important factor to keep in mind in this effort is that a Federal Tax Credit is available which provides a 40% tax credit for expenditures up to $10,000 per year. The economic and other advantages of PV electric systems is discussed quite extensively in Chapter 6.

4.6.2   Example System Design

The best way to illustrate the design process is to do an example problem. We have arbitrarily selected an energy efficient retirement home to be located more than two miles from an existing utility line in the vicinity of Phoenix, Arizona. The house is earth sheltered and oriented to the south so as to minimize cooling requirements and negate the need for all but occasional heating. A solar hot water heater using a PV powered pump is installed to supply nearly all of the hot water requirements.

Since the house is earth sheltered, and the climate is typically dry, evaporative cooling is sufficient to provide comfort on all but the hottest days. Since both the water heater pump, and cooling energy requirements are sun synchronous we can dedicate arrays to those particular uses with no need for battery storage. The arrays for the cooling system can however be wired in such a way that they can provide power for other uses when the cooling system is not needed. Since water heating is needed year round, it is probably better for the water heater array to be isolated from the rest of the system.

Cooking, back-up water heating, and space heating energy are assumed to be supplied by an LP gas system. The rest of the electrical needs are assessed using the suggested table. The four 33 watt (two 1200 CFM LA-9 Systems from Table 3.2) modules dedicated to the cooling system selected can be used as additional back up for periods of occasional cloudiness in the winter months, or to supply Class III (optional) needs when cooling is not required.

Our completed Table 4.1 is shown as Table 4.18. Each of the entries is explained below:

- A portable 50 watt color TV is used for 3 hours in priority class and for 3 hours optional. It is estimated that 50% of the use occurs when the sun is not shining. The storage requirement is then 50% of (150 plus 150) or 150.

- A small Hi-Fi Radio rated at 15 watts is used for 2 hours in priority class and 2 hours optional. Since 50% of use is without sun, storage requirement is 30 watt-hours.

- A fan rated at 25 watts is used 4 hours daily, exclusively during the day so no storage is necessary.

- CB Radio rated at 12 watts is used for 1 hour in priority class and 2 hours options. Fifty percent of use is without sun so storage requirement is 18 watt-hours.

- A small 36 watt refrigerator uses its compressor for 13 hours per day resulting in a Class I (essential) energy need of 480 watt-hours. It is assumed that 50% of the compressor on-time occurs when the sun is not shining, resulting in a storage requirement of 240 watt-hours.

TABLE 4.18. Estimate of PV System Size

Appliance	Power Req.	Daily Energy Requirement			Percent w/o Sun	Storage Req.
		Class I	Class II	Class III		
DC Powered						
Color TV	50		150	150	50	150
Hi-Fi	15		30	30	50	30
Fan	25		100		0	0
CB Radio	12		12	24	50	18
Refrigerator	36	480			50	240
Lighting	20	160			100	160
DC TOTALS		640	292	364		598
AC Powered						
Toaster	1150		288	288	25	144
Wash Mach.	575		173	173	0	0
Vac. Cleaner	690		173	173	0	0
Hair Dryer	1380		173	173	50	173
AC TOTALS			807	807		317

****

Peak Demands (watts): ___1466___   Inverter Capacity: ___1380___
****
Battery Sizing

Storage Req.: ___915___ watt-hours   Depth of Charge Limit: ___10%___

Number of Days Storage: _____   Battery Capacity Req.: __9,150__
****
Array Sizing

Average Daily Load: __1739__ watt-hours   Sun hours per day: ___6___
Storage Req X .2: ___183___ watt-hours   Array Size (watts): ___320___
Adjusted Total: __1922__ watt-hours   Module Req: ___10___
****

- Several 15 watt high efficiency DC lights are used for a total of 8 hours per night. (i.e. two lights for 4 hours each). Since all of this use occurs when the sun is not shining it must all be stored in batteries.

- An AC toaster rated at 1000 watts is used for 15 minutes in priority use and 15 minutes optional. First the power required is increased by 15% to account for the inverter power req. Then this figure is multiplied by 0.25 (1/4 hour) to yield an energy requirement of 287.5 or say 288 watt-hours. 25% of the use is assumed to occur when the sun is not shining resulting in a storage requirement of 144 watt-hours.

- A 500 watt washing machine is assumed to be used exclusively during the daytime for an average of 15 minutes per day in the priority class and 15 minutes optional. Although the typical washing cycle may be longer, this is the estimated time the motor is actually running. There are advantages of running one load daily over 7 loads once a week, because the batteries are more evenly discharged.

- A 600 watt vacuum cleaner is used for 15 minutes daily in priority mode and 15 minutes optional exclusively during the daytime.

- A 1200 watt hair dryer is used for 15 minutes of priority use and 15 optional. Half of the use is expected to occur when the sun is not shining requiring 173 watt-hours to be stored.

Next the columns have been totalled for the DC and AC appliances individually. To estimate power conditioning requirements we assume that the maximum load on the AC systems will occur when the hair dryer is operating alone. Consequently if a 1380 inverter is acquired the user must assure that no other AC appliances are being used at the same time or else a fuse will blow. Another caution is to insure that the surge power requirements of the washing machine and vacuum cleaner do not exceed this figure.

The peak demand is assumed to occur when the hair dryer is used while the color TV is turned on and the refrigerator is running. The figure is the sum of 1380 plus 50 plus 36 or 1466 watts. The size of this number will depend upon the level of control over appliance use which is practicable or desired.

4.41

We then sum the storage requirement of 598 for the DC appliances with the 317 watt-hours for the AC appliances and get 915 watt-hours which is entered in the appropriate space. Since the common rule of 10% depth of discharge is acceptable we determine that a battery system with 10 times the storage requirement is necessary to limit average daily discharge to 10%. This means that 9,150 watt-hours of storage is necessary. Since the weather in Phoenix is generally clear, no additional storage is deemed necessary.

Finally we size the array to supply the system. We add all the power requirements for both AC and DC needs in classes I and II to come up with a average daily load of 1739 watt-hours. This is the sum of 640, 292, and 807. The optional uses are assumed to be exercised only greater than average sun is collected or other loads have been reduced. To this we must add 20% of the storage requirements due to losses in the storage system. The adjusted total energy requirement is 1922 watt-hours per day.

To calculate the capacity of PV array necessary to supply our needs we must consult Figures 4.1 and 4.2 to determine the effective number of sun hours for our location. We use an average figure of 6 for the air-cooler panels that can be diverted for other uses during the least sunny winter months. Dividing our adjusted total by 6 yeilds an array size of 320 watts. If ARCO 16-2000 modules rated at 33 watts each are used, 10 will be required.

Remember that in addition to the 4 panels assumed to be dedicated exclusively to the air coolers we have 1 panel dedicated to the solar water heater pump. Consequently our retirement home would require a total of fifteen 33 watt panels and about 10,000 watt hours of battery storage. Delco 2000 batteries rated at 105 amp hours each will hold 1260 watt-hours apiece (105Ah X 12 volts). Thus 8 batteries would be required.

Connectors and wiring should not be overlooked in the design process. They should be selected after the rest of the design system is complete. A connection diagram should be made before selecting the switches, fuses, circuit breakers and wiring and connectors.

Many of the "off-the-shelf" systems include switches, fuses, connector cables and/or a harness for connecting the system.

One benefit of the photovoltaic power system is the inherent reliability of the solid state components of the system. In any electrical system there are potential maintenance problems. The experienced troubleshooter expects that if all of the components are sound, the most common cause of difficulty would be faulty wire connections. To assure quality system performance, you should:

- Use large enough wires so that overheating does not occur (see the National Electrical Code as well as local codes). These wires become quite large in 12 V systems where substantial quantities of power are to be handled. Thus higher voltages are used as systems become larger.

- Solder the connections or use well designed terminals (Amp Inc. makes a wide variety of wire terminals and one line specifically suitable for PV terminals).

- Select insulation for wires and terminals that is suited to the environment they encounter.

- Protect the wires from future mechanical stress or abuse.

Properly installed wiring and connections add greatly to trouble-free enjoyment of the system.

Using prices listed in Tables 4.7, 4.9 and 4.12, we determine the following system costs:

15	ARCO Solar Modules @ $450 each......................	$6,750
8	Delco Batteries @ $165 each.........................	$1,320
1	Dynamote B-18 Inverter	$ 990
1	UCC Charge Controller...............................	$ 325
TOTAL...		$9,385

The only recurrent cost is replacement batteries approximately every 5 years, meaning that once the system without batteries is paid for at a price of $8,165 the only anticipated operating cost is approximately $264 per year for batteries ($1,320 divided by 5). After taking advantage of the 40% Federal Tax Credit, the net system cost without batteries is less than $5000. How this compares with available alternatives is discussed in Chapter 6.0

4.43

5.0 CUSTOM-BUILT COMPONENTS AND SYSTEMS

The PV industry has the ability to meet most consumer's needs with off-the-shelf components and systems described in Chapters 3.0 and 4.0. However, in those applications where off-the-shelf items will not suffice, several options exist: a) purchase a special order system, b) design and build your own custom system, or c) have someone else design and build you a custom system.

This chapter describes a few of the many custom PV systems already in operation to give the reader an idea of the flexibility and ingenuity of the PV industry. Next, we have compiled a list of PV firms who have expressed the capability of providing custom systems.

5.1 CUSTOM SYSTEMS CURRENTLY IN OPERATION

Currently there are literally thousands of custom PV systems operating in the U.S. Many of the smaller scale systems have been funded through the Federal Photovoltaic Utilization Program, FPUP, (see Table 5.1). The purpose of this program has been to develop the PV technology by establishing demonstration sites that will provide performance information to help direct future R&D efforts.

Other experiments have installed medium and large scale custom systems (see Table 5.2).

The following are a brief description of a few custom systems:

- Solar-powered Warning Lights for
Transmission Towers
Bonneville Power Administration
Washington, Oregon and Idaho.

The Department of Energy and Bonneville Power Administration (BPA) have installed PV warning lights on several transmission towers in Washington, Oregon, and Idaho using Solarex PV arrays. A battery storage system with a capacity of 5000 ampere-hours is also used. This is a more cost-effective system than other approaches. BPA has installed an ampere-hour monitoring device to see if the size of the present system could be reduced.

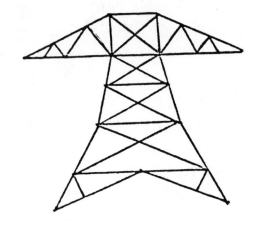

TABLE 5.1 Small Sized Custom Systems Funded Through FPUP

	Peak Power kWp *	Total App.
Residential Sector		
Grid-connected Homes	2-20	2
Individual Homes	1-25	232
Other Residences	0.1-8.5	22
Water Pumps	0.2-10	43
Commercial Sector		
Radiation Samplers	0.04-3	30
Visibility Monitors	1.5	6
Noise Monitors	0.01	2
Navigational Beacons	< 1	18
Weather Monitors	< 1	4
Particulate Sensors	0.04-10.5	26
Meteorological Sensors	< 1	325
Military Ammunition Security		122
Electric Fence	< 1	1
Intrusion Detectors		6
Cathodic Protection		
Pwr Line Towers	< 1	12
Submarine Cables	< 1	12
Transportation Sector		
Beacons	< 0.02-3	1,427
Buoys	< 0.05	3
Anemometer	0.004	1
Moving Target Indicators	1	41
Radar Beacons	< 1	9
Aircraft Arresting Systems	< 1	2
Astronomical Monitor	< 0.5	1
Flash Beacons	0.1-1.5	12
Remote Instrument Platforms	0.5-9	33
Starpex Beacon	< 1	1
Agricultural Sector		
Forest Lookout Towers	< 1	67
Repeaters, Special Purpose	0.01-1.2	35
Venting Sys., Sanitation	< 1	248
Miscellaneous	0.2-25	29
	TOTAL	2,772 Systems

* Where large numbers are installed, the lower figure is more representative.

TABLE 5.2. Intermediate Size Custom Systems

Wilcox Hospital/Acurex, Kauai, Hawaii	35 kW
Sky Harbor Airport/ Arizona Public Service, Phoenix, Arizona	225 kW
BDM Corporation, Albuquerque, New Mexico	50 kW
E Systems, Dallas, Texas	27 kW
Lovington Shopping Center/ Lea Country Electric, Lovington, New Mexico	100 kW
El Paso Electric/ New Mexico State University El Paso, Texas	17 kW
Oklahoma Center for Science and Arts/Science Applications, Inc., Oklahoma City, Oklahoma	135 kW
Beverly High School/Solar Power Beverly, Massachusetts	100 kW
San Bernardino Concrete Plant San Bernardino, California	35 kW
WBNO Radio Station Bryan, Ohio	15 kW
Irrigation & Crop Drying Mead, Nebraska	25 kW
Air Force Station Ft. Belvoir, Virginia	60 kW
College Power System Mississippi Country Community College Blythesville, Arkansas	240 kW
N.W. Mississippi Junior College Senatobia, Mississippi	100 kW
Remote Stand Alone Power System Natural Bridges National Monument Blanding, Utah	100 kW

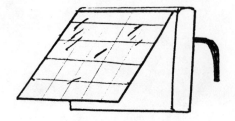

- Solar-powered Back Pack
  Environmental Protection Agency

The Environmental Protection Agency developed a PV powered backpack unit as a power supply for various types of remote instrumentation. The unit uses a Solarex PV panel that can provide up to 120 W. More power can be supplied by interconnecting units. Such a unit could easily be used for other remote applications.

- Solar-powered Indian Village
  Papago Indian Village of Schuchuli
  in Southwestern Arizona

Since 1978, the 15 families of the Papago Indian Village of Schuchuli have had their power needs met by a PV system. The 3.5-kW system provides electricity for a water pump, refrigerators, clothes washer, sewing machine and fluorescent lights. In all cases, energy efficient appliances were chosen.

- Solar-powered Forest Lookout Towers
  Lassen National Forest, California

The U.S. Forest Service has installed a 294 W, 12 V (DC) PV array on two lookout towers. These systems are used to power a 3-cubic foot refrigerator, fluorescent lights, a USFS radio, a bilge pump, and a 12 V (DC) television set. These lookout towers have been in operation since 1976.

5.2 CUSTOM COMPONENTS AND SYSTEMS

Most individual's energy needs will be met with components or systems already designed and built of similar applications. It may take a little time and a few telephone calls, but the chances are, your unique energy need is not unique. You may find that a PV firm designed and built a custom system a few years ago for someone else's unique energy need, and that this firm now offers this system on special order. However, if your energy need can not be met with off-the-shelf or special order systems, many companies within the PV industry will be willing to design and build a custom system. Remember, the design services provided by PV firms are usually not free, and you will be paying more (at least initially) for the custom built system than you would for an off-the-shelf system.

Ask for a cost estimate before any work is done. Then, seriously consider the tradeoffs involved in having a custom system designed and built. Remember, it does pay to shop around!

The following PV companies indicated that they will either design and build custom systems or have systems that can be special ordered (see Table 5.3).

TABLE 5.3. Firms Offering Custom Built or Specialty Components or Systems

Company	Custom Component/System
AAI Corp.	Modules
Abacus	DC to DC converters
Acurex Solar	Concentrating collectors
ARTU	AC Power systems and special DC to AC inverters
Applied Solar Energy Corp.	Concentrating collectors
A. Y. McDonald	Water Pumping systems
Jim Cullen	System design
Roger R. Ethier Assoc.	System design
Free Energy Systems	PV modules and refrigeration, lighting, and communications systems
General Electric	Modules and systems
Globe	Batteries and hardware
Helionetics	Inverters
IOTA Engineering	DC ballasts
March Manufacturing	Pumps
Milton Rov Company	Pumps
Mobil Tyco	Systems
Motorola	Power systems
Parker McCrory	Panels
Photon Power	Arrays
Photovoltaic Energy Systems, Inc.	Consulting & energy information products
Silicon Sensors, Inc.	Arrays
Solar Contractor's, Inc.	PV systems
Solar Power Corp.	PV systems
Solarex	General applications assistance
Solarwest Electric	AC and DC power packages
Solar Energy Corp.	Modules
Solar Usage Now	Remote water pumping systems
Solenergy Corp.	Modules, arrays, PV systems, cells
Spire	PV process equipment, assembling PV modules
Teledyne Inet	Inverters
Tideland Signal	DC power packages
Tri Solar Corporation	Irrigtion systems, home electric systems
United Energy Corp.	Modules and systems
Western Solar Refrigeration	Ice maker, produce cooler, fish freezer, potable water pump
Windworks	Power conditioning equipment
Zomeworks	Passive solar trackers

6.0 WHEN TO CHOOSE PV

The purpose of this chapter is to offer guidance to you in deciding when to select PV powered systems. There are several types of applications where PV has carved out a clear niche and some others where PV doesn't make sense yet. There also is one other type of situation where the wise decision balances on the uniqueness of the location and/or on your individual needs. This latter case generates a more difficult decision for you to make and thus, it will receive a greater emphasis in this chapter.

PV systems are not yet cost-effective in areas near or on a utility grid and for people who wish to continue a normal lifestyle using substantial amounts of electrical power without any serious effort to adjust their electricity use level. Since the concept of a large roof top array generating electricity for domestic needs with some extra to sell to the utility is an exciting vision, there is a substantial program involving private and government funding to some day make it a reality. Although many believe it may soon become practicable, right now the cost of a PV system to do this job would cost about $150,000 per home.

In instances where individuals are willing to accomodate reductions in energy requirements, currently available PV systems may be a viable energy resource. This is especially true if a premium is placed upon the unique benefits of PV systems such as reliability, security, self-sufficiency, environmental compatibility, modularity, and privacy. For those considering building a vacation or retirement home in a remote area, photovoltaic power supplies are well worth serious consideration. For farmers or outdoorsmen, the portability and freedom from utility hook-ups of PV powered systems make them valuable for applications from lighting, to water pumping, to security systems.

There are a growing number of applications where PV systems are clearly preferable to alternative power sources. This is the case where remote power requirements must be met where no utility power is conveniently available for such applications as communications repeaters, navigational aids, and remote sensing equipment. In some instances, the cost of connecting equipment to locally available utility supply may exceed the cost of PV power supplies, a case in point is the frequent use of PV power supplies for aircraft warning lights atop high tension electrical transmission towers. PV systems are also clearly appropriate in instances where power needs are uniquely coincident with solar insolation and the PV supply provides system control as well as power.

The identification of appropriate applications of PV technology requires an assessment of the cost and performance of PV powered <u>systems</u> relative to available alternative <u>systems</u>. The term system is stressed because the optimal method of achieving a specific end may be met through various assemblies of equipment that may vary depending on the power source used. For instance, if water pumping is desired one may choose a mechanical system that directly uses wind to provide pumping, or one may select a PV system that converts sunlight into electricity and then uses electricity to provide pumping.

In the following discussions, we compare the performance of PV systems with appropriate alternative systems where it has particular advantage. The alternatives to be considered include:

- utility power available from the power grid
- wind power for water pumping or electricity generation
- gasoline or diesel generators for electricity
- manual labor

The cost effective application of any of these power supplies necessitates appropriate levels of energy efficiency. The optimal system to fit your needs may be a hybrid that uses two or more of the above energy supplies along with energy conservation principles. Section 6.1 presents applications where currently available PV systems clearly make sense and Section 6.2 discusses instances where PV may be the best power source depending upon your specific needs.

6.1 WHERE PV CLEARLY MAKES SENSE

Photovoltaic energy systems are being employed in a growing number of applications where utility power is either not available or relatively expensive to access. We say relatively expensive, because in some applications the cost of installing a PV powered system is cheaper than connecting a conventional appliance to the utility supply. The applications where PV clearly makes sense can be classified into three categories:

- where power is needed in remote or difficult to access areas
- where net installation costs are lower
- where there is a unique timing of needs

The following discussions highlight examples of these applications and discuss the benefits of PV applications relative to alternative measures.

6.1.1 PV for Remote Areas

PV systems have demonstrated cost-effectiveness in industrial applications away from utility grids where average power requirements are less than one kilowatt. More than 10,000 remote PV applications are installed to power warning beacons on oil drilling platforms, power telephones and repeater stations, radio transmitters on mountain tops as well as for many other applications. These applications have demonstrated superior economics and reliability. Initially these applications were for power requirements in the range of only a few watts per installation. However, with the increasing cost of fossil fuels, and the availability of larger PV systems, larger electrical loads are using PV supplies.

Each installation has unique features to consider in the selection of the best power source. A few of the key attributes of PV power systems are:

1. Reliability - Properly sized PV systems demonstrate excellent reliability in very hostile environments. The costs of failures should be considered.

2. High Altitude Performance PV is favored at high altitudes because the air is thinner and the sun is brighter. For example, at high altitudes a typical diesel generator must be derated to perhaps 1/2 of sea level capacity with consequent losses in efficiency as well as capacity.

3. Maintenance Cost - Transportation of materials and qualified personnel to remote areas for periodic maintenance is very expensive. Since PV systems require only periodic inspection and occasional replacement of storage batteries, maintenance costs are usually much less when compared to alternatives.

4. Cost of Fuel   - Procurement, storage, and transportation costs of fuel
   are unnecessary with PV energy supplies.

Many remote applications have firm or essential power requirements demanding power systems with consistent performance. In such applications alternative energy power systems are frequently unacceptable unless used in conjunction with a backup system. One combination that is being used with success is a hybrid system combining the economy of a wind generator with the consistency of a PV powered system. The overall reliability of the system can be further enhanced by using a petroleum powered backup generator. One clever approach for farmers and construction workers is to use the Lincoln Electric Company product such as their AC-150 WELDANPOWER to provide 115/230VAC power or 150 amperes for arc welding - a handy feature indeed!

For power needs in excess of 1 killowatt continuous, the major alternative in application is diesel for LP gas powered generators. Although diesel systems have relatively low first (purchase) costs, the cost and availability of fuel in remote areas is of such significance that PV arrays are becoming the preferred power source for applications of up to 10 kilowatts. The economics and reliability uncertainties of diesel equipment relative to the security of PV systems are primary influences increasing the use of PV power sources.

In addition to the energy needs for remote industrial equipment, there are significant opportunities for less remote agricultural, residential, and recreational application. A survey of several electric utilities has indicated that the cost for utility line extensions averages over $10,000 per mile. The cost and accesibility of such extensions will vary according to the servicing utility, and the cost will more likely be higher than lower. If soil conditions are less than ideal or if right-of-way for the power lines must be secured, the costs can rise precipitously.

As a consequence, PV supply systems should definitely be considered where utility connection costs are significant. Rather than putting ones money into power lines that must be paid for in addition to the ever increasing price of the power delivered, one can invest in a PV system that has higher reliability and lower operating cost. The farmer who needs to pump water for his livestock may best be served by a PV system that pumps water in relation to the intensity

6.4

of sunlight. Historically, wind systems have been used for this application, and where sufficient wind is available for year round needs it is usually the cheapest alternative. However, in those instances where wind systems have been found to be less than completely sufficient, adding a PV powered pump for a hybrid system may be ideal.

Another example of a remote application would be a vacation home in the wilderness. The alternative sources of power available include wind power, if sufficiently reliable; hydro power, if sufficient and legally useable; a gasoline or diesel generator set, if cost effective and environmentally acceptable; or a silent, reliable, and cost-effective PV system. Once again, if conditions permit, a hybrid system may be the best approach to combine the reliability of a PV system with potential economies of alternatives. Where battery storage is necessary, one of the greatest benefits of hybrid systems is that several types of electrical generation can use the battery storage to its fullest extent at low incremental cost.

6.1.2 Where Net Installation Costs are Lower

PV clearly makes sense in the case of low power needs. Although utility power may be available, local costs for wiring are sometimes sufficient to justify the cost of a PV panel. Retrofitting an attic fan or driveway light can be expensive, however a PV attic fan can be easily installed by the handyman without shock hazard. It is extremely expensive to run a conduit up the side of a petroleum tank to supply a level indicator or a light. It may be more cost efficient to use a small PV array close to the power need for installations such as this.

To assist the reader compare the costs of using a PV system to supply power where connection costs to utility power are significant, we have tabulated typical costs for common wiring tasks in Table 6.1. Actual costs for your installation will vary depending upon the local costs and availability of materials and skilled labor. Although many jobs can be undertaken by "handyman" types, the use of licensed electricians is advisable because of the liability risks of safety hazards resulting from improper appliance installation.

TABLE 6.1. Typical Local Costs of Supplying Remote Branch Circuits

Cost Item	Example Ft.	Total Cost of Wiring
Cost of controls and wiring 120 vac attic fan	75' of Romex	$ 123.48
Same except use 1/2" st'l conduit	75' of Conduit	$ 160.14
Cost of wiring & installing a driveway light	100' Conduit	$ 259.56
Cost of wiring and connecting auto. gate opener on a ranch	500' Conduit	$ 877.80
Cost of wiring a stock watering pump. 1/4 HP 230V	1500' of #8 3 Conductor Cable	$ 2,271.36
Cost of supplying power to a level indicator on petroleum storage tanks	500' to tank, 75' vertical run up tank, 20' horizontal all in Conduit	about $ 5,000

6.1.3 Unique Timing of Needs

PV power supplies also clearly make sense when the energy needs are proportional to the sun's intensity. Such conditions exist in many heating or cooling applications. One example is water circulation pumps in solar heating systems. A circulating pump connected to a PV panel will circulate the water at a rate proportional to the rate of hot water production. It will circulate fastest when the sun is at its hottest.

In such applications, PV offsets the cost of controls normally used and is generally more reliable. The benefits of PV as motive power for solar collection systems include: a) improved reliability, b) improved system efficiency because of proportional control, c) improved protection against freezing because water is not pumped until the sun warms the panel and, d) simplified installation and maintenance. These considerations often outweigh initial cost differences between PV powered and standard controls.

Another example is powering an evaporative cooler with PV panels. In the hottest weather, the sun shines much more than the yearly average, and the PV array produces additional power for the cooling unit. Since demands on utilities are often the greatest during the hottest periods, a growing number of them are adopting pricing systems that charge more for power during these hours. Under these conditions PV powered coolers will have improved cost effectiveness.

The sun-synchronous watering needs of livestock and agricultural areas offer unique opportunities for PV power supplies. The greatest demand for pumping typically occurs when the sun shines the hottest. A PV system is particularly suited to such requirements, and because water storage can replace the need for electrical storage in batteries, system purchase and upkeep costs are minimized.

The key factor in these types of applications is timing. If energy needs are coincident with the solar intensity, PV systems have inherent benefits that often make them the least expensive and most practical form of energy supply. Since these applications use power as it is generated, electrical storage systems are typically unnecessary. Therefore, these systems are among the simplest, most easily maintained and most cost-effective PV applications currently available.

6.2 WHERE PV MAY MAKE SENSE

There is an endless variety of applications where PV may be the most sensible power source depending upon the specific concerns, needs, and preferences of the purchaser. As mentioned in the introduction to this chapter, many of these factors can only be assessed and evaluated by the individual making the purchasing decision. The purpose of this section is to suggest some applications where PV offers particular advantages that may be of value to you.

Although dollar cost is often the most important item in selecting power supplies, many other factors may influence the decision. These include environmental impact, national economy, world security, and feelings of personal well being derived from self-sufficiency. Given the uncertainties regarding the cost and availability of future energy supplies and the stability of the world economy, many of you may place a premium on the certainty a PV energy supply can provide. The fact that PV systems are silent and otherwise environmentally innocuous, simple and easy to maintain, reliable and essentially free of inflationary pressures, may make them attractive at present prices.

Surveys indicate that more people than ever are concerned about our energy future. Since 1975, a significant change has occurred in the way people look at the energy question. According to a recent study (Sumichrast, December 1980, published by National Association of Home Builders), 80% of home purchasers in 1980 considered energy a serious problem whereas in 1975, only about 25% considered energy a problem. Only 7% rated the energy problem in 1975 "extremely serious" but this had increased to about 36% by 1980 and studies show that it may increase to about 74% by 1985. Consequently, a growing number of people are finding that it may be acceptable and even advantageous to reduce their energy dependence through conservation, lifestyle changes, and/or utilization of renewable energy resources.

Clearly there are a large number of people today who are concerned about the future supply and cost of energy. Building a dream home could become a nightmare if utility bills (heating and electrical) aren't controlled. Consequently interest in energy efficient homes and passive solar designs has escalated dramatically in recent years, but surprisingly little attention has been given to the possibility of markedly reducing electrical consumption in the home. If significant reductions in electricity requirements can be accomodated as suggested later in this chapter, PV power systems may prove to be an appropriate and affordable power supply.

Will people continue to buy homes that face an uncertain future of higher fuel bills? Most believe that they will if actions are taken to minimize the cost of energy supplies. This is done first by reducing energy needs through conservation features and lifestyle adjustments, and then procuring dependable and economic fuels. Photovoltaics is the most desireable power supply in many instances today and, as the cost of alternative systems increase, the number of cost effective applications is expected to grow dramatically.

The fact is that the cost of energy from the PV array over its expected 25 year life is constant, and even reduces in real cost compared to the other commodities required for living. The cost of electricity for most utility companies will increase at a rate significantly higher than inflation. On the other hand if a PV system is purchased as part of a 30 year home mortgage, the real burden of those fixed payments typically diminishes as time goes by. If a purchaser's income increases as inflation mounts, and the house payment remains constant, it constitutes a diminishing percentage of total income. The cash flow economics will depend upon your particular perspectives and locale, but it is an important consideration for those nearing retirement, or particularly concerned about their future income potential.

Businesses and residences located in remote areas away from power lines can benefit from PV. Factors to consider in these situations are: a) connection costs, b) costs of extending a power line to your property, and c) advantages of avoiding monthly power bills. If a decision has been made to settle or start a business enterprise where utility power is not readily available, PV may be your best alternative. You must consider the costs of running a utility line to your property (typically in excess of $10,000 per mile), the cost of metering and connection, and finally the cost of power bills.

If a PV system is to be employed to power a building, extensive measures should be taken to reduce the energy needs of the structure. As in other solar systems, it simply does not make sense to use relatively expensive collectors to replace energy that could be more cheaply conserved. The economics of con-servation and PV power supplies is dependent upon conditions at the building site such as climate and energy costs. Although we cannot offer specific guidance for particular applications, we discuss below some example residential installations that reveal the impact of energy conservation and lifestyle adjustments on the size and cost of suitable PV electricity supply systems.

We have conducted the analysis using climatic data for Phoenix, Arizona. The reasons for selecting this location include:

1. The amount of solar insolation (sunshine) is among the highest in the country.
2. The predominant electricity requirement is for space cooling, which is largely coincident with the availability of PV power.

3. The cost of grid supplied electricity is among the highest in the nation.
4. Population growth statistics indicate a continued high rate of home construction.
5. Sufficiently accurate and detailed climatic information is available to use sophisticated energy analysis tools.

Consequently, these examples may represent a "best case" for photovoltaic deployment in the residential sector using currently available PV technonogy.

The analysis consists of four cases conducted in two distinct parts. In the first three cases, we assume that only physical alterations in the design and appliances of the home are made without significant change in the use of the home. We then suggest a case where significant adjustments in lifestyle are made to minimize the cost and maximize the effectiveness of the photovoltaic energy supply system.

In each case a living area of 1,400 square feet has been accomodated, although the design and construction of the home may be significantly altered. The energy supply is sized to meet the typical needs of the average family of three for the first part of the analysis, and is affected only by improvements in appliance efficiency and energy conservation measures. Only the last case assumes that the occupants significantly alter their energy use habits.

It has been assumed that the home <u>is</u> <u>not</u> connected to the utility grid and the PV system is sized to supply all the electricity necessary to:

- keep internal temperatures below 80 degrees F 98% of the time
- maintain adequate illumination levels
- provide power for a refrigerator
- provide adequate power for commonly employed home appliances

Since some of the energy requirement occurs when the sun is not shining, an electrical storage system is required. The cost of this storage requirement has not been rigorously calculated, but is included in the assumed PV system cost of $15 per peak watt. At current costs of $10 per peak watt for PV arrays and $100 per kWh of battery storage, this system cost provides for 20% of power use during non-sunlight periods.

Fuel for cooking, as well as backup for space heating is assumed to be supplied by other means such as LPG, wood, or gasoline powered electric generators. However, a very significant portion of these needs can be met by the PV power supply when excess power is available, particularly if appropriate appliances are selected. Water heating is assumed to be supplied by either electrical or solar energy for the sake of consistency, although a fossil fueled water heater or back-up system would probably be more cost effective considering current PV system and fossil fuel prices. Additional groundrules of the energy use comparisons are listed in Table 6.2.

The energy required for space heating and cooling is dependent upon the characteristics of the building, the comfort requirements of the occupants, and the building climate. We have used the DOE 2.1 computer program developed by the Department of Energy to calculate these requirements. The program uses information about the construction and use of the building in conjuction with weather data for each hour of the typical meterological year to generate estimates of energy needs. The program has been validated against actual building performance in several studies, and has been found to yield estimates within about plus or minus 10%.

In instances where the PV system is the primary source of home electricity, a back-up fossil fueled generator is very useful to provide additional power when extraordinarily low levels of sun, and high levels of need concur. For example, you may occasionally have a large number of overnight visitors increasing energy needs dramatically. It would probably not be cost effective to size the PV array for such short periods of higher than normal need, when a relatively inexpensive fossil fueled generator could be employed for short periods. The cost of this backup system has not been included, however, since the need and value of it vary significantly depending upon the user.

6.2.1 Case I -- Typical All Electric Home

So as to minimize the cost of new homes, energy conservation features are often installed only to the level required by current building codes. Typically, the code is based upon historical rather than projected fuel costs, and

TABLE 6.2. Groundrules of Energy Use Consumption

ARCHITECTURAL AND LIFE STYLE
CHANGES ON THE APPLICATION OF
1981 PV TECHNOLOGY FOR RESIDENCES

PURPOSE

TO ASSESS THE TRADE-OFFS OF ENERGY EFFICIENCY, PV SYSTEM SIZE, AND LIFESTYLE ADJUSTMENT

MEASURES CONSIDERED

INSULATION OF WALLS, CEILINGS, WINDOWS PASSIVE SOLAR, EARTH SHELTERING, WINDOW GLAZINGS, AREAS, AND ORIENTATION, VAPOR BARRIERS, AIR TO AIR HEAT EXCHANGERS, HEAT PUMP, SOLAR WATER HEATING, ENERGY EFFICIENT APPLIANCES AND LIGHTING

METHODOLOGY

COMPARE PV SYSTEM SIZE AND COST WITH CONVENTIONAL ENERGY COST FOR THREE LEVELS OF ENERGY CONSERVATION HOLDING UTILIZATION AND AMENITY LEVEL CONSTANT AND ONE ADDITIONAL CASE WHERE SOME LIFESTYLE AND COMFORT ADJUSTMENTS ARE MADE

ASSUMPTIONS

FAMILY SIZE	3
INTERNAL LOAD	53,100 Btu/day
INFILTRATION	COBLENTZ-ACHENBACH NORMALIZED
PEAK SUN-HOURS	6
SOLAR HOT WATER FRACTION	80
COOLING COP	3.0
ENERGY COST LEVILIZED	8.1¢/kWh
PV SYSTEM COST	$15/WATT
TAX CREDITS	40% FEDERAL UP TO $4000, 35% STATE UP TO $1000
LOCATION	PHOENIX

consequently the amount and cost of energy for heating and cooling is inordinately high. The ranch style home selected as representative of such construction is depicted and described in Table 6.3.

The typical peak electricity needs were found to be approximately 60 kWh per day as shown. The major portion (67%) is for summer cooling, and since solar water heating is not commonly employed, a full 10 kWh per day is required for water heating unless alternative fuels are used. To meet the energy requirements, a full 10 KW of PV capacity is required that would cover nearly one-half of the total roof area.

Even after using the renewable energy tax credits to its fullest extent, the cost of such a large system would be $125,000 and, therefore, well above the realm of affordability by most of us. Thus there is some validity to the widespread public perception that at least at present, photovoltaics are too expensive. However, as you read on, you may realize that this perception is not always justified.

6.2.2 Case II - The Energy Efficient Home

In recognition of the fact that energy costs are expected to continue to escalate as they have in the recent past, numerous buyers are demanding and builders are including additional energy conservation measures in new homes. By incorporating measures that cost no more than $1,375 after tax credits, the average peak energy requirement can be reduced by 40% to approximately 35 kWh per day as shown in Table 6.4.

Items that reduce power consumption at significant additional cost include:

- thicker insulation
- double glazed windows
- complete vapor barrier
- vestibule entry
- use of efficient fluorescent lighting
- solar water heating

None of the cost features incorporated include a two story design that reduces exposed surface area, and concentration of windows on the south facade that maximize solar heating in the winter while minimizing it in the summer.

6.13

TABLE 6.3. Typical All Electric Home

CHARACTERISTICS

INSULATION ——————R-11 WALLS, R-19 CEILINGS

WINDOWS ——————AREA 15% OF FLOOR, SINGLE GLAZING,
EVENLY DISTRIBUTED

INFILTRATION————1.0 AIR CHANGE/HOUR, AVERAGE

ELECTRICAL SYSTEM

		AVERAGE DAILY ENERGY USE
SPACE HEATING	15 kW FORCED AIR FURNACE	35 40*
SPACE COOLING	2 ½ TON FAN COIL	40
WATER HEATING	4.5 kW ELECTRIC RESISTANCE	10
LIGHTING	700 WATTS INCANDESCENT	3
REFRIGERATION	FROST FREE	3
OTHER APPLIANCES	4 kWh/day	4
TOTAL		60 kWh

ENERGY COSTS

$126/month

CONSERVATION MEASURE COSTS

$0

PV SYSTEM SIZE AND COST

60/6 - 10 kW → $150,000

NET SYSTEM COST

THE MAXIMUM TAX CREDIT WOULD BE
$5,000/yr FOR 5 YEARS (1981-1985)
OR $25,000 → $125,000

*SUMMER DESIGN IS MORE SEVERE LOAD

TABLE 6.4. Energy Efficient Home

CHARACTERISTICS

INSULATION —————R-19 WALLS, R-38 CEILINGS

WINDOWS —————AREA 15% OF FLOOR AREA, DOUBLE GLAZED,
PREDOMINANTLY ON SOUTH FACADE

INFILTRATION————0.5 AIR CHANGE/HOUR VESTIBULE ENTRY

ELECTRICAL LOADS

	AVERAGE DAILY ENERGY USE
SPACE HEATING ————10kW FORCED AIRl	25
SPACE COOLING ————1 ½ TON FAN COILl	
WATER HEATING ————80% SOLAR	2
LIGHTING ————50% FLUORESCENT-TASK LIGHTING	2
REFRIGERATION————ENERGY EFFICIENT	2
OTHER APPLIANCES	4
TOTAL	35 kWh

ENERGY COSTS

$74/month

CONSERVATION MEASURE COSTS

INSULATION ———————— $440
WINDOWS ———————— $300
VAPOR BARRIER ————————— $200
SOLAR HOT WATER———————$2500

PV SYSTEM SIZE AND COST

35/6-5.8kW ⟶ $87,500

CREDITS

SYSTEM DOWNSIZING ———————— $200
TAX CREDIT ————————$1875

NET SYSTEMS COST

$87,500 LESS
$25,000 TAX CREDIT + $1365

$63,865

NET

$1365

The cooling system required to condition this home is smaller than typical, and the electricity needs are reduced to 25 kWh on the design day. Since solar water heating is used that adequately meets 80% of the hot water needs, only 2 kWh per day of electricity is needed to heat hot water. Energy needed for the refrigerator is reduced by one-third as a result of selecting a more energy efficient model at an insignificantly higher cost. Although the size of PV systems needed in this case has been reduced to 5.8 KW, the system cost after maximum use of tax credits would be $63,865, still too expensive for most of us.

6.2.3 The Earth Sheltered Home

Many home buyers and builders are experimenting with unconventional construction techniques that offer the promise of dramatically reducing energy needs. The "earth-sheltered" home shown in Table 6.5 is an example that demonstrates significant energy savings can be had if what many consider to be radical architectural practices are acceptable. Earth sheltering reduces energy needs by using the earth as a buffer between the home and the surrounding environment thereby reducing both heating and cooling energy needs. The installation of a continuous vapor barrier is simplified since earth sheltered walls are adequately sealed.

In addition to the features of the home described in Case II this home includes:

- Windows triple glazed and total window area reduced to 10% of floor area.
- Air infiltration has been reduced to a net of .1 air change/hour by using an air to air heat exchanger for ventilation.

This house uses about 1/2 of the electrical energy of the "energy efficient" example because of the reduction on cooling energy requirements that results from earth sheltering and reduction of air infiltration.

The PV system size needed for this home is about 3 KW which would fit nicely on the exposed roof lower level. A system of this size allows the tax credits to be used to their full advantage, covering more that 50% of the total system costs. Although the cost after tax credits is still some $21,125, such a system may be affordable by some, especially when independence, privacy, or self sufficiency is desired.

TABLE 6.5.   Earthsheltered Home

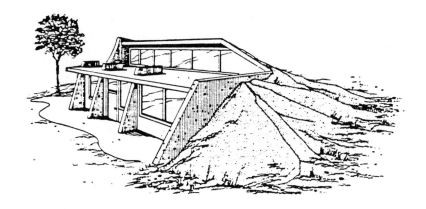

CHARACTERISTICS

INSULATION ———————R-38 CEILINGS
EARTH SHELTERED WALLS

WINDOWS —————————TRIPLE GLAZED SOUTH FACING @ 10% OF FLOOR AREA
OR SINGLE GLAZED WITH PHOTOELECTRIC MOVEABLE
INSULATION

INFILTRATION————————0.1 AIR CHANGE/HOUR NET (0.5 ACH. WITH 75%
EFFICIENT HEAT EXCHANGE) VESTIBULE ENTRY

ELECTRICAL LOADS

	AVERAGE DAILY ENERGY USE
SPACE HEATING	1
SPACE COOLING	8
WATER HEATING	2
LIGHTING	2
REFRIGERATION	2
OTHER APPLIANCES	4
TOTAL	18 kWh

ENERGY COSTS

$38/month

CONSERVATION MEASURE COSTS

EARTH BERMING ———————————$1000
SOLAR HOT WATER ————————————$2500
AIR-AIR HEAT EXCHANGER ——————— $500

PV SYSTEM SIZE AND COST

18/6 - 3 kW ➞ $45,000

CREDITS

SYSTEM DOWNSIZING —————————— $600
INSULATION ———————————————— $400
TAX CREDIT ———————————————————$1875

NET SYSTEM COST

$45,000 LESS
$25,000 BY CREDIT + 1,125

$21,125

NET

$1,125

6.2.4 Case IV -- The Energy Independent Home

For some the desire for energy independence and protection from the uncertainties of our energy future may make significant lifestyle changes and energy use habits acceptable. It is for those who may be willing to implement such changes that we have prepared this example (Table 6.6). Because of the modular nature of PV energy systems, it is possible to start with a small system and add capacity as your needs dictate and finances allow.

It is possible to eliminate 90% of the electrical consumption of a typical home. Naturally, it raises the question if it would be pleasant to live in such a home. Judge for yourself. New types of lighting have been developed that give four times as much light per watt as an incandescent bulb. Refrigerators are available that are smaller than we are accustomed to but operate on 1/4 of the power a standard refrigerator uses. The list goes on.

In addition to the use of efficient appliances, you may consider curtailing unnecessary appliance use such as dishwashers, electric can openers, electric knives etc, and substituting manual methods. Of even greater potential impact is the proper scheduling of energy use activities so as to use power efficiently when it is available in abundance. For example, cooking later in the evening reduces the load on the air conditioner, or washing with full loads only on sunny afternoons, when solar heated water is available in abundance and PV power supply is great, minimizes the size of battery storage.

Can life be fun with 90% reduction in electricity use is always debatable because for you it may be different than for your friend next door. If you could be happy with this kind of situation, you might be able to afford a new kind of pleasure derived from a move toward energy independence. If you could equip your new home with a PV power supply for about $10,000, you would get substantial tax credit and wouldn't have to worry ever again about high electric utility bills ruining the enjoyment of your dream home. Is it a good economic investment? No one can give you the answer to that question because they don't know how high utility bills will go. How much is your peace of mind worth? No one else can decide that for you! It's up to you and no one else.

TABLE 6.6. Energy Independent Home

CHARACTERISTICS

INSULATION ––––––––––R-38 CEILINGS
EARTH SHELTERED WALLS

WINDOWS ––––––––––TRIPLE GLAZED SOUTH FACING @ 10% OF FLOOR AREA
OR SINGLE GLAZED WITH PHOTOELECTRIC MOVEABLE
INSULATION

INFILTRATION––––––––0.1 AIR CHANGE/HOUR NET (0.5 ACH. WITH 75%
EFFICIENT HEAT EXCHANGE)
VESTIBULE ENTRY

ELECTRICAL LOADS

	AVERAGE DAILY ENERGY USE
SPACE HEATING	0
SPACE COOLING	0.8
WATER HEATING	0.2
LIGHTING	0.16
REFRIGERATION	0.5
OTHER APPLIANCES	1.3
TOTAL	2.96 kWh/day

ENERGY COSTS

$7/month

CONSERVATION MEASURE COSTS

EARTH BERMING––––––––––––––$1000
SOLAR HOT WATER––––––––––––$2500
AIR-AIR HEAT EXCHANGER –––––– $500

CREDITS

SYSTEM DOWNSIZING–––––––––– $600
INSULATION –––––––––––––––––– $400
TAX CREDIT ––––––––––––––––––$1875

NET

$1,125

PV SYSTEM SIZE AND COST

2.96/6 = 0.49 $9485

NET SYSTEM COST

$9485 LESS 4794 BY CREDIT
+ 1125 EXTRA HOUSEHOLD COST = $5816

The house features and appearance is identical to the earth sheltered example, however, the use and comfort level of the home are altered somewhat. Specifically this example assumes:

- Evaporative cooling alone is acceptable (the consequence of this assumption is that occasionally the home may be uncomfortably warm for some)
- Efficient fluorescent lights are used for only 8 fixture hrs per day.
- Energy efficient DC appliances are used
- Manual dishwashing and refrigerator defrosting are acceptable
- A smaller 4 cuft energy efficient refrigerator is used
- Back up for water heating is not provided electrically

The example used here is the same as that displayed in Chapter 4.0, Section 4.6.2. It will be recalled that some uses were curtailed. Energy for cooling will not be required at all times and this capacity is made available for other uses when possible.

The cooling load in summer is assumed to be supplied by two evaporative coolers operated by PV panels, power 0.8 Kwhr/day. Hot water is supplied by solar panels with a PV powered circulating pump, power 0.2 Kwhr/day. Lighting uses 12 V DC very high efficiency fluorescent lights and uses 0.16 Kwhr/day. Refrigeration uses a small, high efficiency 12 V DC refrigerator, 0.5 Kwhr/day. Other appliance uses 1.3 Kwhr/day for a total of 2.96 Kwhr/day which is only 1/6 of the energy used in the earth sheltered home without energy management and lifestyle adjustment.

The starter system to provide for these requirements listed in Table 6.6 has a capacity of approximately .5 KW. This size PV system for the Phoenix area has a system cost of about $9,485 using current prices. Tax incentives (Federal and State of Arizona) reduce the net PV cost to less than $5,000. Adding to this the cost of the specified conservation measures of about $1,125 brings the net concept cost of this residence to less than $6,000. This is a very affordable level of expense. If the system was found to be restrictive of life style, it is quite easy to install additional capacity in subsequent years.

The concept may be too restrictive for permanent use but it appears live-able for the first year. You could add to this system in $10,000 increments and receive $5,000 tax benefits each year so that the net cost is greatly reduced. A major advantage of the energy independent concept is that it lets you get started with PV at a modest cost in such a way as to take maximum advantage of the Federal tax credits that are slated to expire in 1985.

6.2.5 Summary of Case Studies

A review of these four case studies is helpful to gain an appreciation of the impact of home energy requirements on photovoltaic system size and cost. Table 6.7 depicts each of the homes with a scaled drawing of the relative size of PV arrays to supply the calculated energy needs. As the annual and peak daily energy use diminishes, the size of the required array drops proportionally.

The typical all electric home consumes approximately 20,000 kWh per year, and requires 60 kWh on a peak day, thereby necessitating a 10 KW array costing about $125,000 even after tax credits. Through the use of conventional conservation measures and passive solar design, the energy needs are reduced by approximately 45% so that only 11,700 kWh per year are needed and the peak requirement is reduced to 35 kWh per day, thereby making a 6 KW array sufficient. Although the conservation features increase the cost of the home by $1,365, the cost of the PV array required is reduced by $60,000.

Clearly one would want to continue adding energy saving measures until their cost exceeds the savings to be had from downsizing the conditioning and photovoltaic systems. Use of earth sheltering principles can cut the energy needs of the energy efficient home again in half, requiring 6,000 kwh per year and only 18 kWh on the peak day. Consequently a 3 KW array is sufficient, costing $20,000 after tax credits. Purchase of a system of this size over a 4 year period allows one to take maximum advantage of federal and state tax incentives, through the purchase of $10,000 worth of arrays per year.

The prior three cases required only insignificant changes in lifestyle resulting solely from the incorporation of energy conservation features on the building and appliances. The final case, called the energy independent home

TABLE 6.7. Comparison of PV Requirements

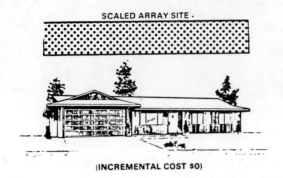

SCALED ARRAY SITE .

(INCREMENTAL COST $0)

(INCREMENTAL COST $1,325)

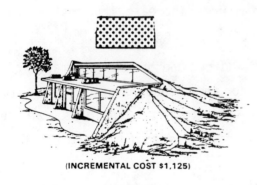

(INCREMENTAL COST $1,125)

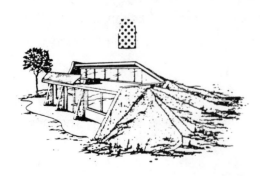

ANNUAL ENERGY USE	PEAK DAILY USE	PV SYSTEM SIZE (kW)	PV SYSTEM NET COST (1981 $)
kWh			
ALL ELECTRIC HOME			
20,000	60	10	125,000
ENERGY EFFICIENCT HOME*			
11,700	35	6	65,000
EARTH SHELTERED HOME*			
6,000	18	3	20,000
ENERGY INDEPENDENT HOME (LIFESTYLE ADJUSTMENT)			
1,080	3	0.5	4,485

*NO LIFESTYLE OR COMFORT ADJUSTMENTS REQUIRED.

6.22

concept, suggests that through prudent alterations in lifestyle to minimize energy needs and managing them in such a way that the effectiveness of the PV array is maximized, the size of the PV array can be drastically reduced. The energy needs of this home are approximatley 1000 kWh per year and the peak daily load is limited to 3 kWh. A .5 KW array is sufficient to supply this level of energy need and would cost less than $5000 after tax credits.

Tax advantages from purchasing larger systems should be considered in system selection, so as to maximize their cost-effectiveness. We present a summary of the federal and state tax credits in Appendix I for more information. We further recommend that you contact your state energy office for more comprehensive and specific guidance in taking advantage of available tax credits.

As mentioned previously the system size can be easily increased in subsequent years because of the modular nature of PV energy systems. Thus a system can be built over a period of years as your needs increase or your finances permit and tax credits will assist you at least until 1985. Since the future of tax credits is uncertain past this period, it is highly advisable to use them now while they are available.

In Table 6.8 we list some preliminary findings of this case study that you may want to consider if you are thinking about purchasing a home electric system. Although these findings are based upon weather data and tax credits of Phoenix Arizona, they are generally applicable to other areas, although optimal system size will likely be different. However, by raising these issues with a local PV supplier, you will be well on your way toward the pruchase of the most cost effective system to meet your particular needs.

6.3 PURCHASING A PV SYSTEM

PV systems have been demonstrated to provide a reliable and cost effective power supply in numerous instances. We have endeavored to highlight some of them in this directory, and as this goes to press, many more are emerging. Because of the rapid expansion of various types and sources of PV products reaching the marketplace, some words of advice to potential purchasers are warranted.

TABLE 6.8. Preliminary Findings

THE EFFECTS OF ENERGY CONSERVATION APPROACHES ON THE ELECTRICITY REQUIREMENTS AND THE POTENTIAL FOR ELECTRICAL ENERGY INDEPENDENCE.

1. PV SYSTEMS ARE MOST COST EFFECTIVELY EMPLOYED ON MORE ENERGY EFFICIENT HOMES

2. UNDER CURRENT FEDERAL TAX REGULATIONS, ONE CAN CLAIM $4000 ANNUALLY FOR THE PERIOD 1981-1985 RESULTING IN A MAXIMUM TAX BENEFIT OF $20,000

3. UNDER ARIZONA STATE TAX REGULATIONS, ONE CAN CLAIM $1000 ANNUALLY FOR THE PERIOD 1981-1985 RESULTING IN A MAXIMUM TAX BENEFIT OF $5000. OTHER STATES HAVE DIFFERENT TAX & INCENTIVE REGULATIONS, WHICH MAY APPLY

4. TAX BENEFIT & CONSEQUENTIAL COST-EFFECTIVENESS IS MAXIMIZED BY MODULARLY BUILDING PV SYSTEMS IN $10,000 INCREMENT ANNUALLY

5. CONSEQUENTLY, THE EARTH SHELTERED STYLE HOME WHICH REQUIRES 3 kW CAPACITY IS OPTIMAL FOR PV DEVELOPMENT WITHOUT CHANGE OF LIFESTYLE. THIS COULD BE APPROACHED WITH A GASOLINE GENERATOR PROGRESSIVELY REPLACED OVER THE PERIOD 1981-1985. PURCHASE OF 3/5kW PER YEAR FOR 5 YEARS

6. ASSUMING GRID EXTENSIONS COST $10,000 MILE, THE EARTH SHELTERED BUILDING IS COST EFFECTIVE MORE THAN 1 MILE FROM THE GRID.

7. WITH APPROPRIATE LIFESTYLE ADJUSTMENT, NECESSARY RESIDENTIAL ELECTRICAL REQUIREMENTS CAN BE SATISFIED WITH A ½kW SYSTEM COSTING LESS THAN $5,000 (AFTER TAX CREDITS). THIS SYSTEM CAN BE UPGRADED IN SUBSEQUENT YEARS, IF DESIRED, TAKING FULL ADVANTAGE OF TAX CREDITS.

Once you have identified a potential application for a PV energy system, contact several vendors to compare the products available. Take the time to explain your needs to the vendor and that you are contacting other suppliers to compare products, services, and warranties. You may want to ask for references of satisfied customers or copies of product brochures and warranties. Most reputable vendors will have little trouble satisfying these requests.

If you anticipate taking advantage of tax credits available for renewable energy systems, contact your state energy office to get the details on their requirements and limitations. They may also maintain lists of licensed vendors or installers as well as individuals using PV energy supplies. In addition, many communities have associations of individuals sharing information on solar energy and energy conservation to which the energy office can refer you for more information.

Industry and government have worked hard to bring this space age technology down to earth so it can be employed to meet our energy needs. The products are available, significant financial incentives are in place, and the risks of failure are very low. The key question for you to ponder is "Am I ready for photovoltaics", rather than "Are photovoltaics ready for me". We stand on the threshold of a secure energy future, and you now have the option to move through it by purchasing a photovoltaic energy system to supply your electrical needs, and set the example for millions to follow.

7.0 USER EXPERIENCES WITH PHOTOVOLTAICS

In addition to reviewing the literature related to the design and performance of photovoltaic power systems, prospective PV purchasers can gain invaluable insight into PV performance from current users. This chapter describes a variety of installations and an appraisal of their performance by the owners. The discussions are based upon information received from personal interviews conducted by the authors.

The characteristics of the PV users interviewed are as diverse as the types of systems described. We interviewed young entrepeneurs, middle aged businessmen, and elderly retirees who use photovoltaics to power mountain retreats, vacation cabins, or year round residences. We talked to wealthy PV users in upper class neighborhoods and exclusive farms as well as young families or retirees using PV to power their modest cabins in remote areas.

This exploration takes you from the sunny beaches of Southern California to the occasionally shadowed hills of Northern California as well as the relatively severe and clouded climate of eastern Washington State just south of the Canadian border. Our intent is to provide information on a sampling of diverse installations in unique locations, purchased by different people for alternate reasons. We hope that these interviews will encourage you to contact PV dealers in your area to view local installations.

In our interviews, we discovered that some people have had problems with their systems. We candidly describe these problems in order to help the prospective purchaser avoid them or at least be prepared to deal with them. Overall we are impressed with the apparent satisfaction of PV owners, whether they have a simple system comprised of one module and a battery and some lights, or a much larger and elaborate system to provide for most domestic needs. There are many other installations (over 1,000) on the west coast of the U.S. that we selected from for the following representative samples.

7.1 SOUTHERN CALIFORNIA INSTALLATIONS

These interviews with southern California homeowners were carried out in June of 1981 through contacts provided by Solarwest Electric Company of Santa

Barbara California and Wm. Lamb Company of Hollywood California. Photovoltaic power systems have proven to be cost-effective in a number of situations where utility power is within 500 yards of the site. In a growing number of areas it is not acceptable to install overhead powerlines, and the need to obtain easements and funds for the installation of underground power connections is sometimes prohibitive.

Currently available PV power systems now make it practical to build homes on excellent view lots regardless of proximity to utility lines. The simplicity and reliability of photovoltaic power supplies provides for ease of installation, no disturbance to the surroundings, and a secure and long lasting energy supply. Although the initial costs of such systems are significant, the freedom from ongoing regulatory and financial requirements associated with utility interconnection frequently override PV system costs.

Doug and Nancy Ingolsby live in the Santa Barbara area with two small children. Their PV system has eight 35 W panels storing energy in about 300 Ah of storage batteries at 24 V DC. They operate lights at 24 V DC using special ballasts, and a 2500 W inverter which supplies power for a stereo systems, a television, a vacuum cleaner, a washing machine and a gas heated clothes dryer. They have sufficient power to do a load of wash daily and meet other typical daily electrical needs. They use gas to power their refrigerator and range.

Because of their remote location, Southern California Edison determined that it was "impossible" to provide power to them. Originally, the Ingolsbys' sole electric supply was a Honda gasoline generator. As they became increasingly concerned about the cost of fuel and the noise associated with its use they began to actively investigate alternative power systems. They ruled out a wind system "because it is mechanical". They report that information on PV was hard to find, but perseverence paid as their satisfaction with their system's performance testifies.

They report that they are very satisfied with their PV systems performance over the past year and feel that a major influence slowing the acceptance of PV power supplies is a lack of helpful information. They were quite pleased to contribute to this effort particularly in light of some of the misinformation being published. They expressed their incredulity when while reading a popular magazine with PV powered light, they came upon a full page advertisment by America's Electric Energy Companies stating that "technology to generate electricity from the sun on a large scale is still a long way from being practical and economical."

John Beaman is a young professional living in a house overlooking the Santa Barbara harbor. His home has been equipped with a four module packaged system obtained from SolarWest Electric Company in 1978. The system provides power for lights, stereo system, television, power tools and an air circulation fan. A 500 watt Triplite invertor provides AC power for the stereo and television. Even though there is no back up system, John reports that he has never been short of power. In the future he plans to power a hot tub pump with the PV power system.

At a year round residence in the foothills of northeast Santa Barbara, Rob and Grace Robinson (see Figure 7.1) present a convincing example of photovoltaic power as a viable alternative energy source. Although the site is only 600 feet from the utility grid the high cost of using utility power prompted the Robinisons to install a solar electric system. Not only does their PV system provide the electricity for their daily needs, it supplied all of the electrical power needed for the home's construction!

FIGURE 7.1. The Robinson Home

7.3

Their system has twelve 35 W modules and a battery bank comprised of twelve Delco 1150 storage batteries connected in such a way as to store power at 24 Volts DC. This power is used directly to provide fluorescent lighting and television reception or is converted either to 12 V DC through a power convertor or to 120 V AC through a 2500 W inverter before delivery to other electrical loads. The house is wired for both DC and AC with each outlet having both an AC and DC socket. This allows for the use of efficient DC lights and the convenience of AC appliances. The inverter is manually controlled by switches located close to the outlets.

The system presently powers 17 interior lights, 3 outside lights, stereo, TV, toaster, blender, juicer, fans, power tools, washer and the dryer motor, vacuum cleaner, and a Jacuzzi pump for an average daily electrical requirement of 1500 watt hours. At this level of consumption the systems storage capacity of 13,000 watt hours could power the home for 5 to 7 days without sunlight. In addition to the PV electricity the homes other energy sources include propane for cooking, refrigeration, and hot water as well as passive solar features and a convection fireplace for space heating.

The home has been occupied full time by two adults since it was completed in August of 1980. The Robinsons report that there has always been ample power and there have been no major problems with the system excepting the repair of the inverter in March which was accomplished under the manufacturers warranty. The payback on the $10,000 system was immediate in light of the tax credit (combined Federal and state) applicable and the $4000 cost of connecting the home to a utility line had the system not been installed.

Rick and Lynn Scott, and their two small children live in a beautiful adobe home in the foothills overlooking Santa Barbara. It became one of the very first homes to use AC from photovoltaic power when the system was installed in 1978 (see Figure 7.2). When the Scott's learned that it would cost $7,500 to bring utility power up 1000 feet to their home, they reasoned the $13,000 cost of a photovoltaic system (less the 55% tax credit) made solar electricity a cost effective alternative.

Their 16 panel system charges twleve Delco 1150 batteries at 48 volts DC. Before the power is used, it is converted to 120 V AC with a 5000 W load

FIGURE 7.2. The Scott Home

demand start inverter. The power is used for 12 interior lights, toaster, stereo, blender, vacuum cleaner, table saw, skil saw, and a juicer for an average of 2,500 watt hours daily. The system provides an average of 3000 watt hours daily and the storage system of approximately 13,000 watt hours can sustain standard building operation for four to five days without sunshine.

This home is similar to the Robinson home in that propane is used for cooking, refrigeration and hot water; and passive solar and wood are used for space heating. The PV system has proven to be a reliable and easily maintained power supply whose overall operation has been very satisfactory. One annoyance that the Scotts revealed is that the load demand start on the inverter does not work for very small loads such as a night light or stereo. Therefore, it is necessary to turn on some other appliance or light to start the inverter.

Jim and Donna Christensen have a 3,300 square foot home in Solvang, California (see Figure 7.3) that was built entirely by PV powered equipment. Although they could now connect to the utility grid, the Christensens choose to enjoy the independence their private electric system provides. They report that "the beautiful part of the PV system is that we are not connected to the utility."

They have twenty seven 35 watt modules connected to charge thirty Delco 2000 batteries at 120 volts DC. A 6 KVA (approx 6000 watt) inverter is

FIGURE 7.3 The Christensen Home

utilized to provide 120 volt AC power for all electrical appliances except for cooking and water heating. This installation is the largest we know of which has been done without government assistance. Except for a few problems which are being corrected, they are very pleased with their system.

The Brisa Del Mar Ranch is one of several in an exclusive limited entry area of the coast near Santa Barbara called the Hollistor Ranch. Dodd Geiger and Gebb Turpin have recently installed a PV powered drip irrigation system (see Figure 7.4) supplying water for 200 avocado trees and 100 citrus trees as well as a large garden there. A 4 Kw Jacobs wind generator supplies electricity for a barn on their ranch with backup supplied by a 6 Kw propane powered Wisconsin generator. Although the wind system performs well it is peridocally necessary to call in outside help to replace the wind turbine brushes since the owners "don't want to climb any 40 foot tower!"

Their irrigation system is powered by twelve 35 W modules connected for 24 V DC. They use about 250 Ah of storage battery to store the early morning sun while the power is too low to operate the progressive cavity pump. The progressive cavity pump, located at 120' depth, is driven by a permanent magnet DC motor on the surface. The pump system by Blue Sky Water Supply Company is

FIGURE 7.4. PV Powered Drip Irrigation System

rated at 2000 gallons per day. The system is currently in a debugging stage, but promises to be a dependable and easily maintained water supply.

William and Sandy Merrill are moving onto a ranch east of Los Angeles where they currently use PV power to pump water and plan to add capacity for domestic lighting and appliance use. Part of their ranch is almost directly under three high voltage power lines (see Figure 7.5). In order to connect to the utility they would have to expend about $4,000 for underground cable, many hours of labor to install it, and obtain an easement from a neighbor who is not cooperating with power companies at this time. They say that PV is an excellent solution to their problem and they like it's quiteness and lack of maintenance.

The water pumping system was installed by William Lamb Company and provides water for livestock, gardening, and domestic use. The installation is termed sun-synchronous since no batteries are used, and water is pumped in direct relation to the available sunlight. The power for the jack pump which lifts 250 gallons per hour from a 230 foot deep well is provided by twelve 35 watt PV modules. Watching their pumping system may not be exciting ("it just keeps on pumping") but without it the land is scorched brown most of the year rather than a productive growth area for produce and livestock.

FIGURE 7.5   TAE Merrill's Pumping System

Although the water system is working perfectly now, there were startup problems which needed to be worked out.   Initially the counterweight was insufficient and the pulleys weren't correctly matched to the motor and pump. It is extremely helpful to have a knowledgeable dealer help with system debugging so as to quickly achieve maximum performance.

The installations discussed above are representative of activity in Southern California.   As this goes to press, many more installations are planned or underway in addition to many which are now in operation but not discussed for the sake of brevity.   If you live in a climate similar to that of the sunny southwestern United States, these experiences should give some insight into the performance of PV power supplies.   We now travel northward so that you can see the uses of PV in climates with less sunshine.

7.2 NORTHERN CALIFORNIA INSTALLATIONS

A wide spectrum of PV system users was encountered in northern California ranging from business and professional people to those espousing an "avante guarde" lifestyle.   Some of the PV owners contacted were hesitant to speak with us because they value privacy very highly; often, this was a motivating factor in their decision to purchase their systems.   We have used pseudonyms to protect those who expressed a concern about it.

The installations described in this section were inspected in June of 1981, and resulted from contacts provided by Jim Cullen Enterprises, Inc. of Laytonville California, Solar General Store of Yarberville California, and Jim Padulla of Padulla Lumber and Co. in Willits California.

Linda McVarish (see Figure 7.6) and Travis Kock live in the Laytonville, California area (P.O. Box 575). They have a hybrid system with four 35 W panels and a wind generator charging a 500 Ah 12 V battery bank. They power 3 fluorescent lights and a dozen incandescent lights at 12 volts and small appliances at 120 V AC using a Tripplite 550 W inverter.

They report that the weakest links in this system are the wind generator ("We may relocate it"), and the inverters. Overall they like the PV system very much and plan to add to it when "the price comes down". They started the backup gas generators about 4 times last winter to charge the batteries during extended cloudy periods, but they expect the generator to "last forever" with the PV and wind system carrying most of the load. Their Servel refrigerator uses 20 gallons of propane in 6 weeks and they are using a gas motor to pump water 100' up a hill to a redwood storage tank. In the future, they plan to use PV to operate the refrigerator and pump water.

Suzanne Rick, Box 323, Laytonville, California uses four 35 W solar modules to power 2 black and white TVs, a stereo, 6 fluorescent, two incandescent lights as well as operate a "cottage industry". She makes original ceramic dolls, and sews authentic costumes for them. The system exceeds her expectations and she would like two more panels ("I use a lot of light"). The system

FIGURE 7.6. The McVarish Home

could also use an elevated adjustable mount for the two panels which are shown in Figure 7.7 mounted on the rafters. The site is less than ideal with trees located all around. Many installations are less than ideally located, but owners are realistic and very pleased with the performance actually obtained.

The James P. Lowary family (P.O. Box 842, Laytonville, California) has ten 18 W panels charging 500 Ah of batteries (Figure 7.8). They operate a color TV, radio, lights, a tape deck, and other 12 V appliances. They have no inverter and operate the AC appliances such as vacuum cleaners from a Honda AC generator powered by propane. They are delighted to eliminate most of the noise and expect that the Honda will last indefinitely with the PV system carrying the major load.

The Lowary's are retired and appreciate controlling their expenses. They have no desire for utility power even if it could be had for less than the $100,000 estimated to extend the power line. They believe the system is not delivering its rated output but they haven't experienced power shortages. They charge the batteries while operating the washing machine, vacuum, etc. with the excess power from the generator. They enjoy the simplicity of PV.

Jerry Martin (P.O. Box 627, Laytonville, California) has six 18 W panels charging 500 Ah of battery storage. He has a 500 W inverter used to power

FIGURE 7.7. The Rick Home

FIGURE 7.8. The Lowary Home

light hand tools. In addition, he powers a TV, stereo, radio, several fluorescent and incandescent lights. He is absolutely sold on PV and plans to add five 35 W panels next month and five more in September. He does not have a regulator on the system and uses a hydrometer to check the batteries once a week. He cleans the battery terminals every two to three months.

Mr. Martin uses the system to provide power to both the trailer and the cabin (see Figure 7.9). He plans to power the refrigerator from PV and perhaps pump the water with it as he enlarges the system.

FIGURE 7.9. The Martin Home

7.11

There are a number of systems in the Garberville, California area whose owners prefer the solitude of the countryside and would not allow their names or pictures of their installation to be published. One of these whom we shall call John Doe has sixteen 35 W panels connected to a 3,500 Ah battery bank that drives a Best 2500 W inverter to supply 120 V AC. A propane generator was used previously and the noise free PV system is most appreciated. He is satisfied with the PV system and would like to enlarge it soon so as to offset the $20/mo propane cost required to fuel his Servel refrigerator.

John Doe II has installed six 35W modules with 700 Ah of battery storage that are protected by a Specialty Concepts regulator. This system is used to power a couple of stereo/radios, a TV set, and numerous fluorescent and incandescent lights in the main cabin as well as a 14' RV. A special application of note is the powering of a DC fan to enhance the effectiveness of a wood stove. He plans to add approximately 8 more panels and an inverter to power a refrigerator and other AC appliances. He was pleased to have less generator noise and looks forward to the larger system that will further reduce propane generator operation.

Other than the replacement of a defective controller in November of 1980, the system has proven to be a 100% reliable power source -- even during the cloudy winter months. The owner feels that <u>such reliable power could not be provided by the utility grid even at the estimated cost of $50,000 for the necessary line extension</u>.

John Doe III built a 1100 S.F. home in a remote area more than ten miles from the nearest power line. Eight months ago he installed four 35 W PV modules to power a variety of 12 V appliances providing music, entertainment, and light. Mr. Doe III is totally satisfied with the system although he did experience a regulator failure which had to be replaced soon after installation. He plans to add four additional panels to power a refrigerator.

We spoke with a customer selecting a module and a 95 Ah battery to power a fluorescent light and a stereo radio. He was pleased to be able to avoid filling the kerosene light in the cabin (particularly during inclement weather) and avoid the odor of kerosene that dominated his environment. Furthermore, without the convenience of a PV power supply the enjoyment of radio broadcasts would be limited by continuous replacement of batteries.

The northern California homeowners we talked to are pleased with their PV systems and confirm that performance is as expected. Most plan to buy more panels "when they can afford it". This allows them to have the immediate convenience that a small system can provide usually for lighting and entertainment systems and the opportunity to gain first hand knowledge of system performance prior to making a large investment.

7.3 <u>NORTHEASTERN WASHINGTON INSTALLATIONS</u>

The installations discussed in this section were inspected in July of 1981 as a result of contacts with the Northeast Washington Appropriate & Creative Technology Organization (NEW ACT). In recognition of the potential value of PV systems to bring the benefits of electrical power to remote residences in the region, the organization sponsored public discussions of photovoltaics and arranged a block purchase in December of 1980 and again in July of 1981. All but one of the systems described below resulted directly from these efforts.

The earliest installation in the region went into operation in 1978 and has provided reliable power for a stereo sound system in a remote mountain retreat (Figure 7.10). One 24 watt Solarex panel is connected to a small aircraft "gel

<u>FIGURE 7.10.</u> PV Powered Mountain Retreat

cell" to power the DC tape player and amplifier. The builder and resident of the retreat home plans to add some small incandescent lights in the near future and comments that her desire for privacy and independence lead her to use photovoltaics. Since the only access to the home is by a steep 500 plus yard trail, lugging batteries, gas generators and fuel, or a wind machine would be difficult. The lightweight photovoltaic panel has proven to be a reliable and trouble free power supply which preserves the silence and pristine naturalness of the site.

The home of Mike Nelson and his wife and teen-aged daughter (Figure 7.11) is an interesting and effective mix of alternative and creative technologies to provide for independent living. The passive solar, partly earth sheltered home employs a small wind generator, two photovoltaic panels, and a 380 amp hour battery storage system to provide power for lighting, home entertainment, and ventilation. Although the cost of connecting the house to nearby power lines is only about $500, Mr. Nelson enjoys the freedom from the requirements and ever escalating costs which utility interconnection would invite.

A look at the history of the home's energy systems provides insight into the appropriateness of the current system. During the construction of the home

FIGURE 7.11. The Nelson Home

a 1,500 watt gasoline generator was purchased to provide the electricity for power tools (Figure 7.12). This same generator provides the large amounts of power for clothes washing and power tools, but is not required for other typical needs. A 200 watt Winco wind turbine mounted atop a 28 ft mast was erected soon after the shell of the house was completed and provided intermittent charging for the 12 volt batteries. The latest addition of two Arco-Solar 35 watt panels in December of 1980 and an inexpensive diode has made the system 100% reliable. Soon Mr. Nelson plans to install a home built 1000 watt wind generator on a 65 foot tower, as well as additional PV modules.

The system powers four 15 watt fluorescent fixtures, two 25 watt incandescent lights, one 25 watt stereo, and a one amp vent fan on a composting toilet. An 50 watt invertor purchased from the J.C. Whitney catalog for $19.95 provides for occasional AC needs such as a turntable. Mr. Nelson states that the maintenance on the system is minimal, only occasional inspection with the hydrometer and topping off of the battery fluid with distilled water. He is completely satisfied with the photovoltaic performance and plans to add more capacity as finances permit to power a small refrigerator.

Dean and Anne Fischer/Lawson have installed a 35 watt panel to provide power for sound and lighting systems. Presently the panel is connected to a

FIGURE 7.12.   1500 Watt Gas Generator

7.15

100 amp hour storage battery which in turn energizes a DC tape deck and amplifier. The home, depicted in Figure 7.13, is more than three miles from the nearest power lines and the only other source of electricity is a rather large 6000 watt Honda gas generator. It would not be cost effective to run this large unit to provide the energy needed for the stereo and small lights. Apparently another panel is planned for the near future since the support frame is already installed.

A local carpenter known as Boone installed a 35 watt panel on his remote residence in January of 1981. The home which is well insulated and includes an attached greenhouse with rock storage to provide both heat and nourishment, is depicted in Figure 7.14. Prior to the PV panel the only electricity available was from storage batteries or a gasoline generator, neither of which was convenient to provide power for lighting and entertainment systems.

The panel is connected to a 100 amp hour snowmobile battery without a voltage regulator. Boone insists that all that is necessary is an occasional topping off of the battery fluid, and that anyone capable of installing an automotive back up light should have no problem with system installation or

FIGURE 7.13 The Fischer/Lawson Home

FIGURE 7.14   The Boone Home

maintenance.   His wife is glad to be free from the inconvenience and danger of kerosene lighting and enjoys the home sound system.   Boone plans to add additional panels to power a refrigerator in the future.

The PV users in this northern climate receive an average of 2 to 4 peak sun hours per day, and report that they are completely satisfied with system performance.   Although their systems are modest they provide a substancial improvement in their lifestyle.   No longer must odorous and expensive kerosene be burned for lighting, and the enjoyment of stereos can be had without the noise of gas generators or the inconvenience and expense of transporting batteries. It is surprising that so much convenience can be had at such a modest cost.

7.4 SUMMARY OF USER EXPERIENCES WITH PV

The PV owners whom we interviewed in California and Washington were quite pleased with the performance of their PV systems.   Occasionally minor problems did arise; however, these were not significant enough to prevent owners from expanding their systems.

Our interviews revealed a fairly consistant pattern of PV system expansion. Most owners begin with 1 to 6 35 W PV modules and a few automobile or golf cart storage batteries. These small systems can be purchased for as little as $500, and will provide power for high priority needs such as music and lighting. Of course, man cannot survive on music and lighting alone, so many users supplement their PV system with a 1750 W Kohler or similar gas generator ($500). This allows the owner to operate vacuum cleaners and other appliances and provides backup to the PV system. Backup lighting is also provided by candles or kerosene.

Once an owner has built-up confidence in a PV system, he enlarges the system by adding a few more panels, batteries, a charge controller and an inverter. This larger system will provide sufficient power for operating most household appliances intermittently except for heating and cooling units and sizeable induction motors. These systems have performed well, and owners seldom use propane or gas generators except in the northern climates.

Some PV owners have enlarged their systems to 16 or more 35 W modules, added more batteries and a larger inverter. This size system is capable of operating all typical AC appliances (except heating and cooling units), minimizing life style changes.

The particular size system that you should purchase depends upon your energy needs, income, location and willingness to accept lifestyle changes. The authors strongly recommend that anyone who is considering a PV system visit a few installations in their area.

APPENDIX I

FINANCIAL INCENTIVES FOR PHOTOVOLTAIC ENERGY SUPPLIES

FINANCIAL INCENTIVES FOR PHOTOVOLTAIC ENERGY SUPPLIES

The Federal government and many state governments offer a variety of financial incentives to stimulate the purchase of renewable energy systems, which usually apply to PV systems. These incentives, if applicable to your installation, significantly affect net purchase cost by allowing the purchaser to:

- exempt the value of the system from property taxes,
- credit part of the installion cost against income taxes, and/or
- be exempt or obtain a refund for sales taxes.

The applicability of these incentives will vary depending on the location and nature of your installation.

As a guide to determine the types of incentives available we have assembled information of Federal and state incentive plants. This information is updated frequently by the National Solar Heating and Cooling Information Center who will provide the information free of charge by calling the numbers listed. Further information can be obtained from your state energy office.

U.S. Department of Housing and Urban Development
Office of Policy Development and Research
In Cooperation with the U.S. Department of Energy

I-2

Solar Data Bank Report

STATE TAX BREAKS FOR RESIDENTIAL SOLAR SYSTEMS

State	Property Tax Exemption	Income Tax Incentive	Sales Tax Exemption
Alabama	no	no	no
Alaska	no	up to $200 credit	not applicable
Arizona	exemption	up to $1000 credit	exemption
Arkansas	no	100% deduction	no
California	yes	up to $3000 credit per application	no
Colorado	exemption	up to $3000 credit	no
Connecticut	local option	not applicable	exemption
Delaware	no	$200 credit for DHW systems	not applicable
Florida	exemption	not applicable	exemption
Georgia	local option	no	refund
Hawaii	exemption	10% credit	no
Idaho	no	100% deduction	no
Illinois	exemption	no	no
Indiana	exemption	up to $3000 credit	no
Iowa	exemption	no	no
Kansas	exemption; refund based on efficiency of system	up to $1500 credit	no
Kentucky	no	no	no
Louisiana	exemption	no	no
Maine	exemption	up to $100 credit	refund
Maryland	exemption statewide plus credit at local option	no	no
Massachusetts	exemption	up to $1000 credit	exemption
Michigan	exemption	up to $1700 credit	exemption
Minnesota	exemption	up to $2000 credit	no

NATIONAL SOLAR HEATING AND COOLING **INFORMATION CENTER**

DR-107
04/80, rev. 03/81
The Center is operated by the Franklin Research Center for the U.S. Department of Housing and Urban Development and the U.S. Department of Energy. The statements and conclusions contained herein are based on information known by the Center at the time of printing. Periodic updates are available. For more information, please contact us at P.O. BOX 1607, ROCKVILLE, MD 20850 or call toll free (800) 523-2929. In Pennsylvania call (800) 462-4983. In Alaska and Hawaii call (800) 523-4700.

STATE TAX BREAKS FOR RESIDENTIAL SOLAR SYSTEMS

State	Property Tax Exemption	Income Tax Incentive	Sales Tax Exemption
Mississippi	no	no	exemption for colleges, junior colleges and universities
Missouri	no	no	no
Montana	exemption	up to $125 credit	not applicable
Nebraska	no	no	refund
Nevada	limited exemption	not applicable	no
New Hampshire	local option	not applicable	not applicable
New Jersey	exemption	no	exemption
New Mexico	no	up to $1000 credit	no
New York	exemption	no	no
N. Carolina	exemption	up to $1000 credit	no
N. Dakota	exemption	5% credit for two years	no
Ohio	exemption	up to $1000 credit	exemption
Oklahoma	no	up to $2800 credit	no
Oregon	exemption	up to $1000 credit	not applicable
Pennsylvania	no	no	no
Rhode Island	exemption	up to $1000 credit	refund
S. Carolina	no	up to $1000 deduction	no
S. Dakota	exemption	not applicable	no
Tennessee	exemption	not applicable	no
Texas	exemption	not applicable	exemption
Utah	no	up to $1000 credit	no
Vermont	local option	up to $1000 credit	no
Virginia	local option	no	no
Washington	exemption	not applicable	no
W. Virginia	no	no	no
Wisconsin	exemption	no*	no
Wyoming	no	not applicable	no

*Wisconsin offers a direct rebate for part of solar expenditures; the rebate is unrelated to taxes.

APPENDIX II

ADDITIONAL SOURCES OF INFORMATION

APPENDIX II

<u>ADDITIONAL SOURCES OF INFORMATION</u>

In this directory, we have tried to give you an idea of the photovoltaic systems/products that are available, and how they can help you meet your energy needs. Although we've raised many questions that you should consider when choosing a system, we haven't raised or answered them all. This section will help you locate the answers to your questions. We have compiled a list of research organizations that can provide you with additional information on government-funded PV installations. Also, we have included a list of books that have proven to be quite helpful in preparing this directory. Many of these books should be available through your public library.

RESEARCH ORGANIZATIONS

Aerospace Corporation
Stanley Leonard
P.O. Box 92957
Los Angeles, CA   95081
(213) 648-7040

Battelle Pacific Northwest Laboratory
Raymond L. Watts
Sigma-4
Box 999
Richland, WA 99352
(509) 376-4348

Jet Propulsion Laboratory
Elmer Christensen
4800 Oak Grove Dr.
Pasadena, CA   91103
(213) 577-9077

Massachusetts Institute of Technology
Lincoln Laboratory
Marv Pope
292 Main Street
Lexington, MA   02173
(617) 862-5500

NASA/Lewis Research Center
Bill Brainard
21000 Brookpark Road
Cleveland, OH   44135
(216) 433-6840

Sandia National Laboratory
Don Schueler
Division 4719
P.O. Box 5800
Albuquerque, NM 87185
(505) 844-4041

Solar Energy Research Institute
Don Ritchie
1617 Cole Blvd.
Golden, CO   80401
(303) 231-1372

OTHER HELPFUL ORGANIZATIONS

- American Section International Solar Energy Society
  American Technological University
  PO Box 1416
  Killeen, TX 76541
  (817) 526-1300

- Solar Energy Industries Association
  George Tenet
  1001 Connecticut Avenue, N.W.
  Suite 800
  Washington, D.C. 20036
  (202) 293-2981

- Renewable Energy Institute
  John Wilson
  1050 17th Street, N.W.
  Suite 1100
  Washington, D.C. 20036
  (202) 822-9157

- Photovoltaics Division, SEIA
  Raymond Kendall, Chairman
  Motorola, Inc.
  8201 East McDowell Road
  Scottsdale, AZ 85257
  (602) 949-3775

REFERENCES

Bechtel National Inc. 1979. <u>Handbook for Battery Energy Storage in</u>
<u>Photovoltaic Power Systems.</u> SAND 80-7022,, San Francisco, CA. The purpose of
this handbook is to provide the photovoltaic system designer with a source of
interface design considerations, as well as performance, cost and other
necessary information involved with the development and design of photovoltaic
systems.

Caskey, D.L., B.C. Caskey and E.A. Aronson. 1980. <u>Parametric Analysis of</u>
<u>Residential Grid Connected Photovoltaic Systems with Storage.</u> SAND 79-2331,
Sandia National Laboratories, Albuquerque, New Mexico. This report investi-
gates the cost of using battery storage in residential grid-connected PV
applications.

Cullen, J. 1980. <u>How to be Your Own Power Company.</u> Van Nostrand Reinhold
Co., New York, New York. This book provides a simple method for designing and
installing your own power system.

Intermediate Technology Development Group. 1980. <u>The Power Guide: A Catalog</u>
<u>of Small Scale Power Equipment.</u> ed. P. Fraenkel, Charles Scribner's Son, New
York, New York. This guide is to help people who want to purchase small scale
power equipment for use in remote and underdeveloped areas. It lists the pros
and cons of choosing various power sources and the availability of these products internationally.

Jacobson, E., G. Fletcher and G. Hein. 1980. <u>Photovoltaic System Costs Using</u>
<u>Local Labor and Materials in Developing Countries.</u> Prepared for NASA Lewis
Research Center by the Engineering Extension Laboratory of the Georgia
Institute of Technology, Atlanta, Georgia. This study addresses the costs of
using photovoltaics in countries that do not presently have high technology
industrial capacity.

Matlin, R. W. and M. T. Katzman. 1977. <u>The Economics of Adopting Solar</u>
<u>Photovoltaic Energy Systems in Irrigation.</u> C0014094-2, Massachusetts Institute
of Technology Lincoln Laboratory. Lexington, Massachusetts. This study compares the costs of PV and conventional fossil fuels as energy sources for irrigation. Also, an estimate is made of the time to initial profitability, and
the time of optimal investment.

Maycock, P., David E. Stirewalt, 1981, <u>Photovoltaics, Sunlight to Electricity</u>
<u>in One Step</u>, Brick House Publishing Co., Andover, MA. This book covers the
story of PV from the unique perspective of a DOE Division Director.
Autographed copies from author, $10, paper, $20, cloth.

NASA Lewis Research Center. 1980. <u>Photovoltaic Stand Alone Systems:</u>
<u>Preliminary Engineering Design Handbook.</u> DOE/NASA/0195-8111, Cleveland, OH.

PRC Energy Analysis Company. 1980. <u>Solar Photovoltaic Applications Seminar.</u>
McLean, VA. This report provided technical information on photovoltaic components and systems.

Paule, T. D. 1981. <u>How to Design an Independent Power System</u>. Best Energy Systems for Tomorrow Inc., Necedah Wisconsin.

Richter, H. P. 1977. <u>Wiring Simplified</u>. Park Publishing, Inc., St. Paul, MN. This book has been written for people who want to learn how to install electrical wiring, so that the finished job will be safe and practical. The installation will comply with the National Electrical Code.

Sandia National Laboratory. 1981. <u>Simplified Design Guide for Estimating Photovoltaic Flat Array and System Performance</u>. SAND 80-7185, Albuquerque, NM. Provides a methodology for estimating PV array and system performance. This report also contains a solar weather data base for 97 U.S. locations.

Solarex Corp. 1980. <u>Guide to Solar Electricity</u>. Washington, D.C. Provides basic information on the operation and use of photovoltaic systems and components. About $7.95 plus handling and postage from Solarex Corp.

Stewart, J. W. 1979. <u>How to Make Your Own Solar Electricity</u>. Tab Books, Blue Ridge Summit, PA. Discusses how PV cells work, the various applications of PV and the future of PV.

APPENDIX III

DOT MATRIX OF MANUFACTURING

DOT MATRIX OF MANUFACTURING

	DISTRIBUTOR/DEALER	HOME ELECTRIC	DC PACKAGE	AC PACKAGE	BATTERY CHARGER	PRODUCTS W/O BAT.	MODULES	BATTERIES	POWER CONDITIONING	DC APPLIANCES	OTHER
AAI CORPORATION			●								●
ABACUS CONTROLS				●					●		
ACUREX CORP.							●				●
AIDCO MAIN CORP.											●
ALPHA ENERGY SYSTEMS	●										
AMETEK, INC.											●
AMP, INC.											●
APPLIED RESEARCH AND TECHNOLOGY				●					●		
APPLIED SOLAR ENERGY CORP.							●		●		
ARCO SOLAR		●	●	●	●	●	●	●	●	●	
A. Y. McDONALD MANUFACTURING			●			●					
BEST ENERGY SYSTEMS									●		
BRADEN WIRE & METAL PRODUCTS, INC.						●					
C AND D BATTERIES					●			●			
CLEAN WATER SYSTEMS										●	●
JIM CULLEN ENTERPRISES, INC.	●										
DSET LABORATORIES, INC.											●
DANFOSS, INC.										●	●
DELCO REMY							●				
DYNAMOTE CORP.									●		
ESB, INCORPORATED								●			
ENERGY HOUSE					●	●	●	●		●	

	DISTRIBUTOR/DEALER	HOME ELECTRIC	DC PACKAGE	AC PACKAGE	BATTERY CHARGER	PRODUCTS W/O BAT.	MODULES	BATTERIES	POWER CONDITIONING	DC APPLIANCES	OTHER
ENERGY UNLIMITED					•					•	
ROGER ETHIER ASSOC.											•
EXIDE CORP.					•						
FREE ENERGY SYSTEMS, INC.			•	•		•		•	•	•	•
GATES ENERGY PRODUCTS									•		
GLASS ENERGY ELECTRONICS	•										
GLOBE-UNION, INC.								•			
HELIONETICS									•		
HYDROCAP CORP.								•			
INTERNATIONAL RECTIFIER							•				
IOTA ENGINEERING, INC.						•					
WILLIAM LAMB CO.			•			•				•	
LI-COR INC.											•
MARCH MFG COMPANY										•	
MID PACIFIC SOLAR						•					
MILTON ROY						•					
MOBIL/TYCO SOLAR						•					
MONEGON LTD.	•										
MOTOROLA INC.						•	•		•	•	
NATURAL POWER, INC.											•
NOVA ELECTRIC								•	•		
PARKER-McCRORY MFG. COMPANY						•					•

	DISTRIBUTOR/DEALER	HOME ELECTRIC	DC PACKAGE	AC PACKAGE	BATTERY CHARGER	PRODUCTS W/O BAT.	MODULES	BATTERIES	POWER CONDITIONING	DC APPLIANCES	OTHER
PENNWALT CORPORATION							●				
PHOTON POWER, INC.							●				
PHOTOVOLTAIC ENERGY SYSTEMS											●
PHOTOVOLTAICS INC.											●
PHOTOWATT INTERNATIONAL, INC.						●	●				
POWER SONIC CORP.								●			
RAIN BIRD INTERNATIONAL						●					
REAL GAS & ELECTRIC CO.						●					
RENEWABLE ENERGY INSTITUTE											●
SES, INC.							●				
SIMICON, INC.											●
SILICON SENSORS, INC.							●				
SILONEX							●				
SOLAR COMPONENTS CORP.	●										
SOLAR CONTRACTORS & BUILDINGS, INC.											●
SOLAR DISC		●									
SOLAR MART	●										
SOLAR POWER CORPORATION							●	●			
SOLAR USAGE NOW, INC.						●					
SOLAR WAREHOUSE											●
SOLAREX CORPORATION	●	●	●	●	●	●	●		●	●	●
SOLARTHERM, INC.						●	●				
SOLARWEST ELECTRIC	●	●	●	●		●	●	●	●	●	●

	DISTRIBUTOR	HOME ELECTRIC	DC PACKAGE	AC PACKAGE	BATTERY CHARGER	PRODUCTS W/O BAT.	MODULES	BATTERIES	POWER CONDITIONING	DC APPLIANCES	OTHER
SOLEC INTERNATIONAL, INC.			●		●		●			●	●
SOLENERGY CORPORATION						●	●	●			
SOLLOS INC.					●		●				
SPIRE CORPORATION							●				●
STANDARD SOLAR COLLECTORS, INC.		●								●	
SURRETTE STORAGE BATTERY CO.								●			
TELEDYNE INET									●		
TIDELAND SIGNAL CORPORATION			●				●				
TOPAZ, INC.									●		
TRI-SOLAR CORPORATION		●	●			●					
UNITED ENERGY CORPORATION					●		●		●	●	
WESTERN SOLAR REF.		●								●	
WILLMORE ELECTRONICS CO.									●		
WINDWORKS									●		
ZOMEWORKS										●	

APPENDIX IV

<u>ADDRESS LIST</u>

AAI Corporation
Nick Kaplan
PO Box 6767
Baltimore, MD 21204
(301) 628-3481

Abacus Controls, Inc.
F. Curtis Lambert
PO Box 893
Somerville, NJ 08876
(201) 526-6010

Acurex
Dariush Rafinejad
485 Clyde Avenue
Mountain View, CA 94042
(415) 964-3200

Advanced Energy Corporation
Gerry Gershenberg
14933 Calvert Street
Van Nuys CA 91411
(213) 728-2191

Aidco Main Corporation
R. Multer
Orr's Island, ME 04066
(207) 833-6700

Alpha Energy Systems
George J. Bauer
120 East Kilgore Road
Kalamazoo, MI 49001
(616) 382-2532

American Power Conversion Corp.
Ervin F. Lyon
89 Cambridge Street
Burlington, MA 01803
(617) 273-1570

Ametek, Inc. AVP Group
Robert A. Russell
1380 Welsh Road
Montgomeryville, PA 18936
(215) 647-2121

AMP, Incorporated
Edgar C. Gorman
Harrisburg, PA 17105
(717) 564-0100

Applied Research and Technology
of Utah (ARTU)
Robert W. McNamara
1918 North 90 West
Orem, UT 84057
(801) 224-2594

Applied Solar Energy Corp.
Thomas J. Brawley
PO Box 1212
City of Industry, CA 91749
(213) 968-6581

ARCO Solar
Photovoltaic Sales & Marketing
Ms. Celine Herzing
20554 Plummer Street
Chatsworth, CA 91311
(213) 988-0667

Arco Solar
Mr. George McClure
20554 Plummer Street
Chatsworth, CA 91311

Automatic Power, Inc.
See Pennwalt

A.Y. McDonald
John D. Eckel
Energy Products Division
12th and Pine Street
Dubuque, IA 52001
(319) 583-7311, Ext. 227

Battelle
Pacific Northwest Laboratories
Raymond L. Watts
Sigma IV Bldg.
P.O. Box 999
Richland, WA 99352
(509) 376-4348

Best Energy Systems
Mr. Terrance D. Paul
Route 1, PO Box 106
Necedah, WI 54646
(608) 565-7200

Blue Sky Water Supply Co.
Ronald Shaw
PO Box 21359
Billings, MT 59104
(406) 259-0654

Braden Wire & Metal Products, Inc.
Martin L. Steger
PO Box 5087
San Antonio, TX 78201
(512) 734-5189

C & D Batteries
R. N. Kauffman
Market Development Asst.
3043 Walton Road
Plymouth Meeting, PA 19462
(215) 828-9000, Ext 246

Clean Water Systems
Jack Cotter
15155 Stagg Street
Van Nuys, CA 91405
(213) 782-1207

Jim Cullen Enterprises, Inc.
The Wilderness Home Power System
PO Box 732
Laytonville, CA 95454
(707) 984-6186

DSET Laboratories, Inc.
Ms. P.V. French
Box 1850, Black Canyon Stage
Phoenix, AZ 85029
(602) 465-7356

Danfoss, Inc.
Stephen M. Madigan
16 McKee Drive
Mahwah, NJ 07430
(201) 529-4900

Delco Remy
D.L. "Sonny" Williams
2401 Columbus Avenue
Anderson, IN 46011
(317) 646-7404

Dynamote Corp.
Stephen Handley
1200 West Nickerson
Seattle, WA 98119
(206) 282-1000

ESB Incorporated
2510 North Boulevard
Raleigh, NC 27604
(919) 834-8465

Energy House
Jack Zohar
PO Box 848
Mercer Island, WA 98040
(206) 236-0282

Energy Unlimited, Inc
Ted Gurniak
#2 Aldwyn Center
Villanova, PA 19085
(215) 525-5215

Roger Ethier Assoc.
205 Franklin Street
Alexander, VA 22314
(703) 683-2657

Exide Corp. (Subsidiary of ESB)
Mr. Gene Cook
101 Gibraltar Road
Horsham, PA 19044
(215) 441-7482

Free Energy Systems, Inc.
Al Bakewell
Holmes Industrial Park
Holmes, PA 19043
(215) 583-4780

Gates Energy Products
Mike Harrison
1050 S. Broadway
Denver, CO 80217
(303) 744-4806

General Electric
Mr. Jim Marler
PO Box 8661, Room 114
Philadelphia, PA 19101
(215) 962-5835

Glass Energy Electronics
Ron Wilson
4463 Woodland Park, Ave. N.
Seattle, WA 98103
(206) 632-1645

Globe-Union, Inc.
Battery Division
Department G
Paul C. Bronesky
5757 North Green Bay Avenue
Milwaukee, WI 53201
(414) 228-2581

Helionetics
Delta Electronic Control Division
Stan Boyle
17312 Eastman Street
Irvine, CA 92714
(714) 546-4731

Honeywell Motor Products
Roger Baird
PO Box 106
Rockford, IL 61105
(815) 966-3600

Hydrocap Corp.
George Peroni
975 N.W. 95th Street
PO Box 380698
Miami, FL 33150
(305) 696-2504

International Rectifier
Harold Weinstein
233 Kansas Street
El Segundo, CA 90245
(213) 322-3331

Iota Engineering, Inc.
Sylvia D. Clayton
1735 E. Ft. Lowell Rd., Suite 12
Tuscon, AZ 85719
(602) 327-5781

Kohler Company
44 High Street
Kohler, WI

William Lamb Co.
10615 Chandler Blvd.
North Hollywood, CA 91601
(213) 980-6911

LI-COR, Inc.
Photovoltaic Product Marketing
John Gewecke
PO Box 4425
Lincoln, NB 68504
(402) 467-3576

March MFG Company
Mr. F. Ahline
1819 Pickwick Avenue
Glenview, IL 60025
(312) 729-5300

Massachusetts Institute of
   Technology, Lincoln Laboratory
Marv Pope
PO Box 73
Lexington, MA 02173

Mid-Pacific Solar
Roberta Colvin
15808 Arminta St.
Van Nuys, CA 91406
(213) 908-0655

Hartell Div Milton Roy
Mr. Douglas Bingler
70 Industrial Drive
Ivyland, PA 18974
(215) 322-0730

Mobil/Tyco Solar
Mr. K. V. Ravi
16 Hickory Drive
Waltham, MA 02154
(617) 890-0909

Monegon LTD.
Mr. Harold L. Macomber
4 Professional Drive
Suite 130
Gaithersburg, MD 20760
(301) 840-0320

Solavolt International
Mr. Raymond Kendall
Solar Energy Dept.
5005 East McDowell Road A110
Phoenix, AZ 85008
(602) 244-6513

NASA Lewis Research Center
Bill Brainard
21000 Brookpark Road
Cleveland, OH 44135

Silonex Inc.
Dr. M. Pawlowski
Chief Engineer, Silicon Devices
331 Cornelia Street
Plattsburgh, NY 12901
(518) 561-3160

Natural Power, Inc.
Brian Gordan
Francestown Turnpike
New Boston, NH 03070
(603) 487-5512

Norcold Incorporated
R. C. Matz
1510 Michigan Street
Sidney, OH 45365
(513) 492-1111

Nova Electric Man. Co.
Kenneth Niovitch
263 Hillside Ave.
Nutley, NJ 07110
(201) 661-3434

Pacific Energy Systems
Edward Delvers
427 E. Montecito Street
Santa Barbara, CA 93101
(805) 963-2155

Parmak
Parker-McCrory MFG. Company
Kenneth D. Turner
3175 Terrace Street
Kansas City, MO 64111
(816) 753-3175

Pennwalt Corporation
Automatic Power Division
Robert Dodge
Hutchinson Street
Houston, TX 77002
(713) 228-5208

Photon Power, Inc.
Mr. Martin Wenzler
13 Founder Blvd.
El Paso, TX 79906
(915) 779-7774

Photovoltaic Energy Systems, Inc.
Paul Maycock
2401 Childs Lane
Alexandria, VA 22308
(703) 780-7308

Photovoltaics, Inc.
Mr. Lawrence Curtin
1110 Brickle Avenue
Suite 430
Miami, FL 33131
(305) 374-2440

Photowatt International, Inc.
George W. Lacey Jr.
2414 West 14th Street
Tempe, AZ 85281
(602) 894-9564

Power Sonic Corp.
Bruno A. Ender
Photovoltaic Marketing
PO Box 5242
3106 Spring St.
Redwood City, CA 94063
(415) 364-5001

Rain Bird International, Inc.
Kenneth A. Ude
7045 North Grand Avenue
Glendora, CA 91740
(213) 963-9311

Real Gas & Electric Co.
James Weller
PO Box F
Santa Rosa, CA 95402
(707) 526-3400

Renewable Energy Institute
John Wilson
1050 17th Street, N.W., Suite 1100
Washington, D.C. 20036
(202) 822-9157

SES, Incorporated
Greg T. Love
Tralee Industrial Park
Newark, DE 19711
(302) 731-0990

Sandia Laboratories
Don Schueler
Division 4719
PO Box 5800
Albuquerque, NM 87115
(505) 844-4041

Semicon, Inc.
Robert A. Irvin
14 North Avenue
Burlington, MA 01803
(617) 272-9015

Silicon Sensors, Inc.
Ronald W. Ignatius
Highway 18 East
Dodgeville, WI 53533
(608) 935-2707

Silonex, Inc.
331 Cornelia St.
Plattsburgh, NY 12901
(518) 561-3160

Solar Components Corp.
Mr. Scott Keller
PO Box 237
121 Valley St.
Manchester, NH 03105
(603) 668-8186

Solar Contractors & Buildings, Inc.
Mr. Michael Baker
8 Charles Plaza #805
Baltimore, MD 21201
(301) 727-6740

Sun Force
Genos Colvin
540 Lagoon Drive
Honolulu, HI 96819
(808) 833-0001

Solar Electronics
156 Drakes Lane
Summertown, TN 38483
(615) 964-2222

Solar Energy Research Institute
Don Ritchie
1617 Cole Boulevard
Golden, CO 80401
(303) 231-1372

The Solar Mart
Division of McGill Stevens, Inc.
134 Vermont N.E.
Albuquerque, NM 87108

Solar Power Corporation
(Affiliate of Exxon Corp.)
C. William Clark
20 Cabot Road
Woburn, MA 01801
(617) 935-4600

Solar Usage Now, Inc.
Mr. Joseph Deahl
Box 306
420 E. Tiffin Street
Bascam, OH 44809
(419) 937-2226

Solar Warehouse
Mr. Richard Brandt
140 Shrewsbury Ave
Red Bank, NJ 07701

Solarex Corporation
Ed Roberton
Marketing
1335 Piccard Drive
Rockville, MD 20850
(301) 948-0202

Solar General Store, Inc.
Bob Walker
445 Conger Lane
Garberville, CA 95440
(707) 923-3353

Solartherm, Inc.
Carl Schleicher
1110 Fidler Lane
Silver Springs, MD 20910
(202) 882-4000

Solarwest Electric
Rob Robinson
232 Anacapa Street
Santa Barbara, CA 93101
(805) 963-9667

Solec International, Inc.
Robert F. Brown
12533 Chadron Avenue
Hawthorne, CA 90250
(213) 970-0065

Solenergy Corporation
Mr. Robert Willis
171 Merrimac St.
Woburn, MA 01801
(617) 938-0563

Sollos, Inc.
Dr. Milo Macha
2231 Carmelina Avenue
Los Angeles, CA 90064
(213) 820-5181

Specialty Concepts
820 N. Lincoln Street
Burbank, CA 91506

Spire Corporation
Mr. Thomas Wilber
Patriots Park
Bedford, MA 01730
(617) 275-6000, Ext. 223

Standard Solar Collectors, Inc.
Claude Burly
1465 Gates Avenue
Brooklyn, NY 11385
(212) 456-1882

Surrette Storage Battery Co.
PO Box 3027
Salem, MA 01970
(617) 745-4444

Teledyne Inet
Fred Tamjidi
2750 West Lomita Blvd.
Torrance, CA 90509
(213) 325-5040

Tideland Signal Corporation
Carl Kotila
4310 Directors Row, Box 52430
Houston, TX 77052
(713) 681-6101

TriSolarCorp
Mr. Ron Matlin
10 DeAngelo Drive
Bedford, MA 01730
(617) 275-1200

Topaz, Inc.
Powermark Division
Chuck Parker
3855 Ruffin Road
San Diego, CA 92123
(714) 565-8363

United Energy Corporation
Bill Edwards
1176-D Aster Avenue
Sunnyvale, CA 94086
(408) 243-0330, or
PO Box 31089
Honolulu, HI 96820
(808) 836-1593

Western Solar Refrigeration, Inc.
Ronald L. Strathman
715 "J" Street
San Diego, CA 92101
(714) 235-6002

Willmore Electronics Co.
Photovoltaics Products
Chris Ely
PO Box 1329
Hillsborough, NC 27278

Windworks
Thomas Werking
Box 44A, Route 3
Mukawongo, WI 53149
(414) 363-4088

Wisco Divison, ESB Incorporated
Mr. Jerry Fox
2510 North Boulevard
Raleigh, NC 27604
(919) 834-8465

Zomeworks
PO Box 712
Albuquerque, NM 87103
(505) 242-5354

Zond
427 E. Montecito
Santa Barbara, CA 93101
(805) 966-4167

INDEX